The Philosopher's Tree

Michael Faraday by Thomas Phillips

The Philosopher's Tree

A Selection of Michael Faraday's Writings

compiled with commentary by

Peter Day, FRS

Fullerian Professor of Chemistry
The Royal Institution of Great Britain

CRC Press
Taylor & Francis Group
Boca Raton London New York

CRC Press is an imprint of the
Taylor & Francis Group, an **informa** business

First published 1999 by IOP Publishing Ltd.

Published 2019 by CRC Press
Taylor & Francis Group
6000 Broken Sound Parkway NW, Suite 300
Boca Raton, FL 33487-2742

First issued in paperback 2019

No claim to original U.S. Government works

ISBN 13: 978-0-367-44759-5 (pbk)
ISBN 13: 978-0-7503-0570-9 (hbk)

Visit the Taylor & Francis Web site at
http://www.taylorandfrancis.com

and the CRC Press Web site at
http://www.crcpress.com

British Library Cataloguing-in-Publication Data
A catalogue record for this book is available from the British Library.

Library of Congress Cataloging-in-Publication Data are available

Typeset in TEX using the IOP Bookmaker Macros

May Providence grant you health and strength, so that you may *very very long* continue an Honor to old England, and a splendid living Example that, by calling to his aid industry the most indomitable, perseverance the most unflinching, and honesty the most unsullied, a poor uneducated Lad may by God's blessing make himself, not only one of the first philosophers of Europe, but also one of the most Enviable Characters of the Age.

J South to M Faraday
30 December 1844

It is my wish, if possible, to become acquainted with a method by which I may write...in a more natural and easy progression. I would, if possible, imitate a tree in its progression from roots to a trunk, to branches, twigs and leaves, where every alteration is made with so much ease and yet effect that, though the manner is constantly varied, the effect is precise and determined.

M Faraday, letter to B Abbott
31 December 1816

Contents

Contents

Introduction

Playing the game of picking winners is a decidedly twentieth-century pastime: composers, poets, novelists, footballers, all have been subjected to orders of merit and ranked in leagues, or perhaps only the top eleven or hundred. So why not play the game with scientists, whose work (after all) has given us the world frame for modern life? The first eleven begins to pick itself: Galileo, Newton, Darwin, Rutherford and Faraday, but the first four transformed the way we view the natural world each in quite specific, although vitally important, respects. The name Faraday stands apart because his discoveries and insights extended across such a wide canvas, from organic chemistry and electrochemistry through the science of materials to electromagnetic induction, the universal property of diamagnetism and, perhaps most fundamental of all, the connection of magnetism to light, which opened the door to the unification of the natural forces and the concept of fields. And all was grounded in observation and beautifully conceived and executed experiments of a startling simplicity and directness. Given this circumstance, the number of books expounding every detail of Faraday's life and work is legion. Is there, therefore, any need to add to their number and can there, indeed, be any aspect of Faraday's genius, and of the world he inhabited, that remains unexplored?

Successive Directors of the Royal Institution, where Faraday lived and worked almost continuously from 1813 to 1862, are peculiarly susceptible to the spell of this extraordinary man. It is simply not possible to inhabit the second floor flat where he and his wife

Sarah carried on their everyday lives, and walk through the parts
of the building that he knew, without being aware of that benign
but determined presence; or to hold the small pieces of home-made
apparatus (original archetypes of the phrase 'string and sealing
wax') that constitute the world's first transformer and dynamo
without awe that one person could achieve so much. And it is this
presence that leads one back to answer the question just posed.

Faraday's social origins, his thought processes, his methods
of experimentation and his religion have all been subjects of
exhaustive analysis by historians and philosophers of science.
One aspect of his work, which gives unique insight into the
path by which his career developed, and the way in which his
mind worked, appears not to have received much emphasis
outside the realm of the academic professionals: namely, his
writing. Throughout his life, from the time when he was a teenage
apprentice bookbinder until his final resignation from the Royal
Institution due to failing memory, Faraday wrote voluminously
and his output took many forms. Before telegraph, phone and fax
machine (not to mention electronic mail) correspondence was the
only viable way to communicate at a distance. Apart from letters,
though, Faraday kept journals (both scientific and personal);
as a practising scientist, he wrote articles in learned journals;
as an adviser to Government and to many other agencies, he
wrote reports; as a supremely successful communicator (especially
to young people), he left lecture notes and transcripts.   All
these add life, colour and depth of focus to the stereotypical
scientific colossus voyaging (though Wordsworth's phrase applies
to Newton) 'through strange seas of thought alone'.   Though
Faraday's life was largely lived within what might appear to be
very narrow geographical confines (just a few miles around 21
Albemarle Street in London's West End) his professional, social
and family relationships were extensive and diverse, and his
responses to them equally complex; and through all the forms
of expression that his multifaceted career required of him, one fact
shines clearly. Not only is he in the world first eleven (or fewer)
for science; he shows enviable quality as a writer.

Scientists whose writings are vivid and approachable may not be the norm for their profession. (Neither, come to that, are the historians or literary critics.) However, in the example of scientists, they form a distinguished lineage, from Darwin to Feynman. Within this tradition, however, Faraday's written output deserves special mention for two reasons. First is the variety of forms it encompasses; from the most spontaneous and intimate (whether it be to his teenage friend and confidant Benjamin Abbott or the self-communing of his laboratory notebook) to the most resounding public statements such as his letters to the *Times* and his Christmas lectures to young people. Second are the remarkable circumstances of his upbringing. The story of how the young bookbinder's apprentice, and son of an indigent blacksmith, came to take up a career in research, and became one of the foremost savants of his (among others) age, is one of the greatest romances of science, indeed, a story so remarkable that it would scarcely be credited as an invention. That story has been told so many times that it would be superfluous, indeed wearisome, to repeat it. The general bibliography at the end of this book lists a short selection of biographies. Rather, the purpose of the present anthology is to allow Michael Faraday to come alive for a wider audience (as he did in his 'discourses' and young peoples' lectures at the Royal Institution) through his own words.

The present anthology makes no pretence of being a scholarly work. It contains only the minimum of commentary necessary to situate the context in which the various extracts were written, and how they relate to the broad sweep of Faraday's life and work. It is arranged by theme: personal, social, professional and to a certain extent chronological. Consisting, by its nature, of extracts, it can make no claim to be comprehensive: Faraday wrote 158 learned articles in scientific journals! It is a personal selection, emphasising passages that seem to me to come closest to revealing the man. In my own teenage years, a Shakespeare anthology, *The Ages of Man*, compiled by the distinguished critic George Ryland opened a window into the vast domain of that giant's thought and expression. I hope the following pages may serve in a similar way to introduce others to the self-taught former bookbinder's apprentice who was also (I say it boldly) the greatest experimental physical scientist in the history of the world.

A Note about Sources

Comparatively little of Faraday's writing remains unpublished except for some of his letters which are still being collected for the monumental *Correspondence of Michael Faraday*, in course of publication by my colleague Dr Frank James [1]. I am extremely grateful to Dr James, not just for allowing me to plunder the results of his exhaustive scholarship in the cause of popularisation, but for sharing with me many of the insights he has gained into Faraday's personality and methods of working through the task of editing his correspondence. Given Faraday's fame as a scientist, and the widespread interest in his life and achievements, the number of biographies looking at every aspect of his activities is legion. Among the earliest, and having the advantage of being written from personal knowledge, is that of Bence Jones, for many years Honorary Secretary of the Royal Institution [2]. In these two magisterial volumes were reproduced for the first time many of Faraday's letters and other unpublished material. In modern times the most authoritative biography, placing Faraday's chemistry and physics in a wider historical and philosophical context, is that of Pearce Williams [3]. The same author also edited two volumes of selected correspondence which, until Dr James' work is finished, remains the most complete collection of letters covering the later part of Faraday's career [4]. The experimental diary was published verbatim in seven volumes under the editorship of Thomas Martin in the 1930s [5], but whilst it had been well known since Bence Jones published extracts from it, Faraday's journal of his continental tour between 1813 and 1815 was not published verbatim until 1991 (the year of his bicentenary) under the editorship of Brian Bowers and Lenore Symons [6].

Books compiled from the transcripts of Faraday's Christmas Lectures on *The Chemical History of a Candle* have been in print continuously since the lectures were first given in 1848–1849 [7], though, curiously, the book arising from his other published lectures on *The Various Forces of Nature* [8] has been out of print for over a century.

In conclusion, it may be thought odd that I have not included extracts from some of the 158 scientific articles in which Faraday presented his experiments and conclusions in learned journals, mostly the *Philosophical Transactions of the Royal Society* and the *Philosophical Magazine* [9]. For this deliberate omission there are several reasons. First, my aim in this book is to collect examples of Faraday's writing at its most direct and spontaneous: formal articles in what scientists call 'archival' journals are rarely notable for their living prose, and Faraday's are no exception. Second, the arguments that they develop, by way of sifting and evaluating evidence, are often complicated and difficult to do justice to through extracts. Finally, in an effort to emphasise this towering figure's humanity, my emphasis in choosing pieces to reproduce veers towards relationships, with his family, his friends and colleagues and (in a quite personal way) with his science. What this book contains is a subjective and infinitesimal sampling of a gigantic treasure: the books cited above (as well as others in the Bibliography) will provide more insight.

A Note about Punctuation

Whether vivid and spontaneous, or magisterial and declamatory, Faraday's prose style is a remarkable product for one whose formal schooling finished at the age of 12. An Achilles heel (not, in that age, confined to autodidacts alone) is his punctuation. Quite rightly, from a historical point of view, the definitive compilations of his letters, notebooks and scientific papers present Faraday's writing exactly as it appears on the original written manuscripts where, in some circumstances, no full stop or comma appears for pages at a time. Since the present volume is not intended for professional historians or philosophers, but for the interested public, I have taken the liberty of including punctuation where appropriate to improve readability.

Chapter 1

The Beginning: In a Nutshell

The story of how Michael Faraday, born the son of a blacksmith on 22 September 1791 and later apprenticed to a bookbinder, came to be one of history's greatest scientists, is so remarkable that in a work of fiction it could be taken as pure fantasy. Let us therefore begin this account of Faraday's life and work in his own words

The facade of the Royal Institution after the addition of columns in 1838. Watercolour drawing by T Hosmar Shepherd.

The house where Faraday grew up in Jacob's Well Mews.

with the brief outline of his first encounter with Humphry Davy, and arrival at the Royal Institution, where he was to spend the remainder of his working life. Just after Davy's death in 1829, when he had already been at the Royal Institution for sixteen years, he replied to an enquiry from John Ayrton Paris, the first biographer of Humphry Davy, as to how he first met Davy.

When I was a bookseller's apprentice, I was very fond of experiment and very averse to trade. It happened that a gentleman, a member of the Royal Institution, took me to hear some of Sir H. Davy's last lectures in Albemarle Street. I took notes, and afterwards wrote them out more fairly in a quarto volume.

My desire to escape from trade, which I thought vicious and selfish, and to enter into the service of Science, which

I imagined made its pursuers amiable and liberal, induced me at last to take the bold and simple step of writing to Sir H. Davy, expressing my wishes, and a hope that, if an opportunity came in his way, he would favour my views. At the same time, I sent the notes I had taken of his lectures. Early in 1813 he requested to see me, and told me of the situation of assistant in the laboratory of the Royal Institution, then just vacant.

At the same time that he thus gratified my desires as to scientific employment, he still advised me not to give up the prospects I had before me, telling me that Science was a harsh mistress and, in a pecuniary point of view, but poorly rewarding those who devoted themselves to her service. He smiled at my notion of the superior moral feelings of philosophic men, and said he would leave me to the experience of a few years to set me right on that matter.

Finally, through his good efforts I went to the Royal Institution early in March of 1813, as assistant in the laboratory and, in October of the same year, went with him abroad as his assistant in experiment and in writing. I returned with him in April 1815, resumed my station in the Royal Institution, and have, as you know, ever since remained there.

Chapter 2

Early Years: Friends, Family and Marriage

While young Michael was working in Mr Riebau's book shop in Blandford Street, just off Baker Street in the centre of London, he had a wide circle of friends, based on the City Philosophical Society. This group was a good deal less formal than the name

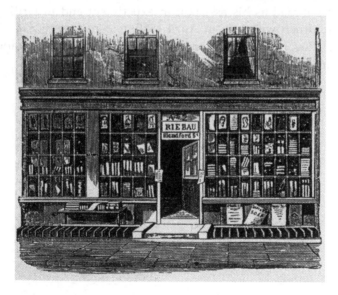

Mr Riebau's bookshop in Blandford Street.

suggests but it provided a forum for ideas, or just friendship, as the following extract from a letter in 1811 to a fellow member, Thomas Huxtable, shows:

> Tit for tat, says the proverb; and it is my earnest wish to make that proverb good in two instances. First you favoured me with a note a short time since, and I hereby return the compliment; and, secondly, I shall call 'tit' upon you next Sunday, and hope you will come to tea 'tat' with me the Sunday after. In short, the object of this note is to obtain your company, if agreeable to your convenience and health (which I hope is perfectly recovered long before this), the Sunday after next.

Youthful Friendship: Benjamin Abbott

Michael Faraday's closest friend, encountered through the City Philosophical Society, was Benjamin Abbott. Not only were they frequent visitors to one another's homes but, even though they were living only a mile or two apart in central London, they exchanged letters. What first brought them together was a mutual interest in chemical experiments. For example, Faraday exultantly describes the electrolytic cell he made from simple pieces bought in a shop with his meagre pocket money, and how he used it to decompose a solution of magnesium sulphate:

> I have lately made a few simple galvanic experiments merely to illustrate to my self the first principle of science. I was going to Knights to obtain some Nickle, & bethought me that they had Malleable Zinc. I enquired & bought some. Have you seen any yet? The first portion I obtained was in the thinnest pieces possible; observe it in a flattened state. It was, as they informed me, thin enough for the Electric Snake, or as I before called it, de Luc's Electric column. I obtained it for the purpose of forming discs, with which & copper to make a little battery. The first I completed contained the immense number of seven pairs of Plates!!!

and of the immense size of half-pennies each! I, Sir, covered them with seven half-pence and I interposed between seven or rather six pieces of paper, soaked in a solution of Muriate of Soda!!!—but to laugh no longer Dear A _ _ _ _, rather wonder at the effects this trivial power produced. It was sufficient to produce the decomposition of the Sulphate of Magnesia; an effect which extremely surprised me, for I did not, (I could not) have any idea that the agent was component to the purpose. A thought has struck me—I will tell you, I made the communication between the top & bottom of the pile & the solution with copper wire: do you conceive that it was the copper that decomposed the earthy sulphate? (That part, I mean, immersed in the solution) that a galvanic effect took place I am sure, for both wires became covered in a short time with bubbles of some gas, & a continued stream of very minute bubbles, appearing like small particles, rose through the solution from the negative wire. My proof that the Sulphate was decomposed was that in about 2 hours the clear solution became turbid; Magnesia was suspended in it.

After Faraday started work at the Royal Institution as the Chemical Assistant, he still wrote to Benjamin Abbott, sometimes about experiments that were going on, but also to set down in a quite personal way his developing opinions about the methods of science, about the nature of scientific explanations but beyond them all, the importance of human dialogue and friendship. Here he is describing how he got home in the rain after a Sunday gathering on 19 July 1812, thinking in a youthful spontaneous way as he went, about all manner of scientific issues. Exuberance leaps from the page:

July 20, 1812

Monday Evening. 10 o'clock

Dear Abbott,

Were you to see me, instead of hearing from me, I conceive that one question would be how did you get home on

Sunday evening? I suppose this question because I wish to let you know how much I congratulate myself upon the very pleasant walk (or rather succession of walks, runs and hops) I had home that evening, and the truly Philosophical reflections they gave rise to.

I set off from you at a run and did not stop until I found myself in the midst of a puddle and quandary of thoughts respecting the heat generated in animal bodies by exercise. The puddle, however, gave a turn to the affair and I proceeded from thence deeply immersed in thoughts respecting the resistance of fluids to bodies precipitated into them. I did not at that time forget the instances you and your brother had noticed in the afternoon to that purpose.

My mind was deeply engaged on this subject, and was proceeding to place itself as fast as possible in the midst of confusion, when it was suddenly called to take care of the body by a very cordial, affectionate & also effectual salute from a spout. This of course gave a new turn to my ideas and from thence to Blackfriars Bridge it was busily bothered amongst projectiles and parabolas. At the Bridge the wind came in my face and directed my attention as well as earnestly as it could go to the inclination of the pavement. Inclined planes were then all the go and a further illustration of this point took place on the other side of the Bridge, where I happened to proceed in a very smooth, soft, and equable manner for the space of three or four feet. This movement, which is vulgarly called slipping, introduced the subject of friction, and the best method of lessening it, and in this frame of mind I went on with little or no interruption for some time except occasional and actual experiments connected with the subject in hand, or rather in head.

The velocity and momentum of falling bodies next struck not only my mind but my head, my ears, my hands, my back and various other parts of my body, and tho I had at hand no apparatus by which I could ascertain those points exactly, I knew that it must be considerable by the quickness with which it penetrated my coat and other parts of my dress. This happened in Holborn and from thence I went home sky-gazing and earnestly looking out for every Cirrus, Cumulus, Stratus, Cirro-Cumuli, Cirro-Strata and Nimbus that came from above the Horizon.

But chemistry was more to the fore in young Michael's mind than physics at that time. In September 1812 we find him explaining that the reaction between phosphorus chloride and ammonia is not an acid–base reaction of what we would nowadays call the Brønsted type, but something quite different, leading to the conclusion that chlorine is an element (what he called a 'simple body'):

The substance (the Chloride of Phosphorus, as we will call it) combines with Ammonia—here I fancy you crying out 'an Acid, an Acid, it combines with an Alkali'!!! but softly my good Sir, we have no acid as yet. Tho it does combine with Ammonia, no Phosphate is formed but a dry powder. This powder is very different to the combination of PA & Ammonia and possesses different properties & characters. It is exceedingly fixed in the fire; it will not rise at a white heat whereas the Phosphate of Ammonia is decomposed at an heat far below that point. Consequently it must be a different substance & Chlorine must be a simple body.

In the spring of 1813 he was able to describe to Abbott how Sir Humphry Davy met with a very nasty accident on preparing a new explosive substance that we know now as nitrogen trichloride. This compound had been prepared first by the French chemist Dulong, who lost an eye and several fingers when handling it. Experimental chemistry in Davy's Laboratory, too, was a hazardous business:

You desire me to inform you at times of any thing new in philosophy that may fall in my way and I shall accordingly obey your desires by detailing to you at present some circumstances relating to the newly discovered detonating compound. This I do with more eagerness as I have been engaged this afternoon in assisting Sr. H. in his experiments on it, during which we had two or three unexpected explosions.

This compound is formed by inverting over a solution of the Nitrate or Muriate of Ammonia an air jar full of fresh made pure clean chlorine gas, all contact of oil, grease or inflammable matter being carefully guarded against. It was at first supposed necessary to surround the solution with ice but it is of no importance to do so; it in fact forms better without it.

Immediately that the gas is inverted over the solution an action commences. This is evident by the gradual tho slow rise of the solution in the jar. As the absorption of the gas takes place quickly, spots are evident on the surface of the solution in the jar which increase in size and appear as drops of an high coloured oil. As the action goes on, these drops become so large as at last to fall from the surface and sink to the bottom of the solution.

With respect to its detonation powers, it exhibits them with many bodies when a small portion of it is placed in a basin and covered with water, and oil or Phosphorus is then brought in contact with it. It explodes violently; the basin is shattered to pieces and the water is thrown in all directions. But I can inform you of a very easy and safe method of inflaming it by oil, which is thus: drop a small portion of it on an oily surface and an instantaneous inflammation will ensue but without noise. Heat also explodes this body and it was by this means that Sr. H. met with his very unpleasant accident.

Faraday himself had some narrow escapes, as another letter to Abbott a few days later reveals:

Agreeable to what I have said above, I shall at this time proceed to acquaint you with the results of some more experiments on the detonating compound of Chlorine and Azote, and I am happy to say I do it at my ease, for I have escaped (not quite unhurt) from four different and strong explosions of the substance. Of these, the most terrible was when I was holding between my thumb and finger a small tube containing about 7 and a half grains of it. My face was within twelve inches of the tube but I fortunately had on a glass mask. It exploded by the slight heat of a small piece of cement that touched the glass above half an inch from the substance and on the outside. The expansion was so rapid as to blow my hand open, tear off a part of one nail and has made my fingers so sore that I cannot yet use them easily. The pieces of tube were projected with such force as to cut the glass face of the mask I had on.

But then the correspondence ranged far beyond accounts of chemical accidents and theories. How scientific (or philosophical, as they called it) argumentation worked and how theories were arrived at (or abandoned) were the subjects of repeated exchanges. Do we detect a hint of mild exasperation on Faraday's part with his young friend's apparent unwillingness to tackle more than one issue at a time, or to abandon positions when they became untenable in the face of young Michael's highly intellectual onslaught?   Two letters of August and September 1812 are revealing:

I have to notice here a very singular circumstance, namely a slight dissent in my ideas from you. It is this: you propose not to start one query until the other is resolved, or, at least, 'discussed and experimented upon' but this I shall hardly allow, for the following reasons: ideas and thoughts often spring up in my mind and are again irrevocably lost for want of noting at the time. I fancy it is the same with you and would therefore wish to have any such objections or unresolved points exactly as they appear to you in their full force, that is, immediately after you have first thought of them. For to delay until the subject in hand is exhausted would be to loose all the intervening ideas. Understand, too, that I preserve your communications as a repository into which I can dip for a subject requiring explanation, and therefore the more you insert, the more it will deserve that name. Nevertheless I do not mean to desert one subject for another directly it is started, but reserve them for after subjects of consideration.

Definitions, dear A., are valuable things. I like them very much and will be glad, when you meet with clever ones, if you will transcribe them. I am exceedingly well pleased with Dr. Thomson's definition of Chemistry. He calls it the science of insensible motions: 'Chemistry is that Science which treats of those events or changes in natural bodies which consist of insensible motions' in contradistinction to Mechanics which treats of sensible motions.

How do you define idleness?

'Read it through'

You wrong me, dear Abbott, if you suppose I think you obstinate for coinciding in my opinion immediately. On the contrary, I conceive it to be but proper retention. I should be sorry indeed were you to give up your opinion without being conceived in error in it, and should consider it a mark of fickleness in you that I did not expect. It is not for me to affirm that I am right & you are wrong; speaking impartially, I can as well say that I am wrong and you are right, or that we both are wrong and a third right. I am not so self opinionated as to suppose that my judgement & perception in this, or in other matters, is better or clearer than that of other persons, nor do I mean to affirm that this true theory is reality but only—that my judgement conceives it to be so. Judgements sometimes oppose each other, as in this case, and as there cannot be two opposing facts in nature, so there cannot be two opposing truths in the intellectual world. Consequently when judgements oppose, one must be wrong—one must be false—and mine may be so for aught I can tell. I am not of a superior nature to estimate exactly the strength & correctness of my own and other men's understanding and will assure you, dear A., that I am far from being convinced that my own is always right.

I have given you this theory, not as the true one, but as the one which appears true to me, and when I perceive the errors in it I will immediately renounce it in part or wholly as my judgement may direct. From this, dear Friend, you will see that I am very open to conviction but from the manner in which I shall answer your letter you will also perceive that I must be convinced before I renounce.

Beyond their mutual interest in experiments and theories it is clear that Faraday's friendship with Abbott was close and deep. Nevertheless Faraday remained aware of how much he wanted to accomplish and how limited was the time that he could devote to each part of his busy life; as we see from his letter to Abbott of 2 August 1812:

Dear Abbott,

What is the longest, and the shortest thing in the world: the swiftest and the most slow: the most divisible and the most extended: the least valued and the most regretted: without which nothing can be done: which devours all that is small: and gives life and spirits to everything that is great?

It is that, Good Sir, the want of which has till now delayed my answer to your welcome letter. It is what the Creator has thought of such value as never to bestow on us mortals two of the minutest portions of it all at once. It is that which with me is at the instant very pleasingly employed. It is Time.

How joyful was the spark of intellectual curiosity and debate that passed between the two friends shines out from Faraday's reply written on 1 October 1812 to a letter from Abbott of 28 September:

—no—no—no—no—none—right—no   Philosophy   is

not dead yet — no — O no — he knows it — thank you — 'tis impossible — Bravo.

In the above lines, dear Abbott, you have full and explicit answers to the first page of yours dated Septr 28. I was paper hanging at the time I received it but what a change of thoughts it occasioned; what a concussion, confusion, conglomeration, what a revolution of ideas it produced. Oh 'twas too much. Away went cloths, shears, paper paste and brush, all. All was too little, all was too light, to keep my thoughts from soaring high, connected close with thine.

Two further letters from Faraday to his friend, of October 1812 and May 1813, expand on his feelings for Abbott and on the primacy of personal friendship, especially its moral dimension:

I would much rather engage the good opinion of one moral philosopher who acts upon his precepts than the attentions and common place friendship of fifty natural philosophers. This being my mind, I cannot fail to think more honorably of my Friend since the confirmation of my good opinions, & I now feel somewhat satisfied that they are correct.

I thank that cause to whom thanks is due that I am not in general a profuse master of those blessings which are bestowed on me as a human being. I mean health, sensation, time & temporal resources. Understand me clearly here for I wish much not to be mistaken. I am well aware of my own nature: it is evil and I feel its influence strongly. I know too that—but I find I am passing insensibly to a point of divinity, and as those matters are not to be treated lightly I will refrain from pursuing it. All I meant to say on that point was that I keep regular hours, enter not intentionally into pleasures productive of evil, reverence those who require reverence from me and act up to what the world calls good. I appear moral and hope that I am so, tho' at the same time I consider morality only as a lamentably deficient state.

I have made use of the term 'friend' several times and in one place I find the expression of common place friendship. It will perhaps not be improper at this time to give you my idea on true friendship and eligible companions. In every action of our lives I conceive that reference ought to be had in a superior being and in nothing ought we to oppose or act contrary to his precepts.

According to what I have said a few lines above, I would define a friend (a true friend) to be 'One who will serve his companion next to God'. Now will I admit that an immoral person can fill completely the character of a true friend. Having this idea of Friendship, it was natural for me to make a self inquiry whether I could fill the character, but I am not yet satisfied with my own conclusions on that point. I fear I cannot. True friendship I consider as one of the sublimest feelings that the human mind is capable of, and requires a mind of almost infinite strength and at the same time of complete self-knowledge. Such being the case, and knowing my own deficiency in those points, I must admire it, but fear I cannot attain it.

The second takes as its starting point the fact that, amidst all the preoccupations filling his life he had given no thought to his friend, nor received any letter from him for some time. The reference to the music playing from a hotel at the rear of the Royal Institution leavens the moralising:

The monk, for the chastisement of his body, and mortification of sensual lusts and worldly appetites, abstains from pleasures and even the simple supplies that nature calls for. The miser for reasons as strong, tho' diametrically opposite (the gratification of a darling passion), does exactly the same and leaves unenjoyed every comfort of life. But I, for no reason at all, have neglected that which constitutes one of my greatest pleasures and one that may be enjoyed with the greatest propriety—'till on a sudden, as the dense *light* of the electric flash pervades t'horizon, so struck the thought of Abbott through my soul.

And yet, Ben, 'tho I mean to write to you at this time I have no subject in particular out of which I can cut a letter. I shall therefore (if you will allow me a second simile) follow the pattern of the expert sempstress who, when she has cut out all her large and important stuff, collects and combines as fancy may direct pieces of all sorts, sizes, shapes and colours and calls it patch work. Such a thing will this epistle most probably turn out. Begun one day, yet most likely finished on another; formed of things not otherwise connected than as they stand upon the paper (things, too, of different kinds). It may well be called patch work or work which pleases none more than the maker.

What is the matter with the thumb and forefinger of your right hand? And yet 'tho they be ever so much out of order, it can scarcely excuse your long silence. I have expected something from you before now even 'tho it be wrote thus.

'He hath not music in his heart &c'—confound the music say I. It turns my thoughts quite round (or rather half way round) from the letter. You must know, Sir, that there is a grand party dinner at Jaques hotel which immediately faces the back of the Institution, and the music is so excellent that I cannot for the life of me help running at every new piece they play at the window to hear them. I shall do no good at this letter tonight and so will get to bed and 'listen, listen to the voices of' bassoons, violins, clarinets, trumpets, serpents and all the other accessories to good music. I can't stop. Good-night.

What a singular compound is man—what strange contradictory ingredients enter into his composition and how completely each one predominates for a time according as it is favoured by the tone of the mind and senses and other existing circumstances. At one time, grave circumspect & cautious; at another silly, headstrong and careless; now conscious of his dignity, he considers himself as Lord of the creation. Yet in a few hours will conduct himself in a way that places him beneath the level of the beasts. At times free, frivolous and open, his tongue is an unobstructed conveyor of his thoughts: thoughts which, after consideration, make him ashamed of his former behaviour. Indeed, the numerous paradoxes, anomalies and contradictions in man exceed in number all that can be found in nature elsewhere, and separate and distinguish him (if nothing else did) from every other created object, organised or not. The study of these circumstances is not uninteresting, in as much as knowledge of them enable us to conduct ourselves with much more propriety in every situation in life. Without knowing how far we ourselves are affected by them, we should be unable to trust to our discretion amongst other persons, and without some knowledge of the part they bear or make in the composition, we should be unable to behave to him unreserved & with freedom.

It was my intention when I again sat down to this letter to obliterate all the former part of it but the thoughts I have just set down were sufficient to alter my determination. I have left them as being the free utterance of an unemployed mind and delineating a true part of my constitution. I believe too, that I know sufficiently of the component parts of my friend as to justify my confidence in letting them remain unaltered.

After his year and a half absence on the continent with Sir Humphry Davy and Lady Davy (see Chapter 3), Faraday's meetings with Abbott, and letters to him grew less frequent. Still, from time to time, he called on his old friend for advice, as in the following, where he turns a note of apology for not having kept up with their correspondence into a request for tips on how to cultivate a spontaneous style of writing: something about which, from the examples given here, it may be thought that he was little in need of instruction:

I must confess that I have always found myself unable to arrange a subject as I go on, as I perceive many others apparently do. Thus I could not begin a letter to you on the best methods of renovating our correspondence and,

Benjamin Abbott in later life.

proceeding regularly with my subject, consider each part in order and finish (by a proper conclusion) my paper and matter together. I always find myself obliged (if my argument is of the least importance) to draw up a plan of it on paper and fill in the parts by recalling them to mind, either by association or otherwise and, this done, I have a series of major & minor heads in order & from which I work out my matter. Now this method unfortunately (though it will do very well for the mere purpose of arrangement & so forth) introduces a dryness and stiffness into the style of the piece composed by it, for the parts come together like bricks, one flat on the other and, though they may fit, yet have the appearance of too much regularity and it is my wish, if possible, to become acquainted with a method by which I may write my exercise in a more natural and easy progression. I would, if possible, imitate a tree in its progression from roots to a trunk, to branches, twigs & leaves, where every alteration is made with so much ease & yet effect that, though the manner is constantly varied, the effect is precise and determined. Now in this situation I apply to you for assistance.

The transformation in the young Michael's life that was brought about when he joined the Royal Institution is alluded to in a characteristically understated way, in a letter to an aunt and uncle. The impending absence that he refers to is the visit to the continent, described in Chapter 3:

London, Sept. 13th, 1813.

As for myself, I am absent (from home) nearly day and night except occasional calls, and it is likely shall shortly be absent entirely, but this to you. I was formerly a bookseller and binder, but am now turned philosopher, which happened thus: Whilst an apprentice, I, for amusement, learnt a little of chemistry and other parts of philosophy, and felt an eager desire to proceed in that way further. After being a journeyman for six months, under a disagreeable master, I gave up my business, and, by the interest of Sir H. Davy, filled the situation of chemical assistant to the Royal Institution of Great Britain, in which office I now remain, and where I am constantly engaged in observing the works of Nature and tracing the manner in which she directs the arrangement and order of the world.

Marriage to Sarah Barnard

Faraday met his future wife Sarah through the small nonconformist religious sect, the Sandemanians, to which he, like his father and many of his family, belonged. Several letters speak eloquently of his feelings and, on one occasion at least, he tried his hand at verse. The lines alluded to have vanished, but Faraday seems to have been stricken by remorse at what he had written: apparently commending cool reason as superior to warm emotion:

R.I.

Oct. 11th 1819

You ask'd me last night for the lines which I penn'd,
    When, exulting in ignorance, tempted by pride,
I dared torpid hearts and cold breasts to commend,
    And affection's kind pow'r and soft joys to deride.

If you urge it I cannot refuse your request:
    Though to grant it will punish severely my crime:
But my fault I repent, and my errors detest;
    And I hoped to have shown my conversion in time.

Remember, our laws in their mercy decide
    That no culprit be forced to give proof of his deed:
They protect him though fall'n, his failings they hide,
    And enable the wretch from his crimes to reced.

The principle's noble! I need not urge long
    Its adoption; then turn from a judge to a friend.
Do not ask for the proof that I once acted wrong,
    But direct me and guide me the way to amend.

M.F.

Some ten months later, in July 1820, Faraday was still asking Sarah to exert her benign influence on his character, and not withdraw her friendship:

You know me as well or better than I do myself. You know my former prejudices, and my present thoughts—you know my weaknesses, my vanity, my whole mind; you have converted me from one erroneous way, let me hope you will attempt to correct what others are wrong.

Again and again I attempt to say what I feel, but I cannot. Let me, however, claim not to be the selfish being that wishes to bend your affections for his own sake only. In whatever way I

Sarah Barnard: a pencil sketch.

can best minister to your happiness, either by assiduity or by
absence, it shall be done. Do not injure me by withdrawing
your friendship, or punish me for aiming to be more than a
friend by making me less; and if you cannot grant me more,
leave me what I possess, but hear me.

Still, with so much of busy professional life to occupy (and
exhaust) him, Faraday continued to write to Sarah, even when he
had little to say:

Royal Institution: Tuesday evening

My dear Sarah,—It is astonishing how much the state of the
body influences the powers of the mind. I have been thinking
all morning of the very delightful and interesting letter I
would send to you this evening, and now I am so tired, and
yet have so much to do, that my thoughts are quite giddy,
and run round your image without any power of themselves
to stop and admire it. I want to say a thousand kind and,

believe me, heartfelt things to you, but am not master of words for the purpose; and still, as I ponder and think on you, chlorides, trials, oil, Davy, steel, miscellanea, mercury, and fifty other professional fancies swim before and drive me further and further into the quandary of stupidness.

From your affectionate

Michael

They were married on 12 June 1821, a union which (though childless) was to last to the end of Faraday's life. The joy of his life with Sarah shines through a letter he wrote to her on 21 July 1822 from Paternoster Row in London, where he was staying while Sarah was away visiting her family:

And now, my dear girl, I must set business aside. I am tired of the dull detail of things, and want to talk of love to you; and surely there can be no circumstances under which I can have more right. The theme was a delightful and cheerful one before we were married, but it is doubly so now. I now can speak not of my own heart only, but of both our hearts. I now speak, not only with any doubt of the state of your thoughts, but with the fullest conviction that they answer to my own. All that I can now say warm and animated to you, I know that you would say to me again. The excess of pleasure which I feel in knowing you mine is doubled by the consciousness that you feel equal joy in knowing me yours. Oh, my dear Sarah, poets may strive to describe and artists to delineate the happiness which is felt by two hearts truly and mutually loving each other; but it is beyond their efforts, and beyond the thoughts and conceptions of anyone who has not felt it. I have felt it and do feel it, but neither I nor any other man can describe it; nor is it necessary. We are happy, and our God has blessed us with a thousand causes why we should be so. Adieu for to-night . . .

Both the Faraday and Barnard families remained close and there were frequent visits to one another's houses, as well as letters

Sarah Faraday: a photograph taken in the sitting room of the flat at the Royal Institution, 1841.

exchanged about the ordinary moments of life. Sarah and Michael took several brief holidays at Brighton, principally to give Michael a rest from the unremitting labour of laboratory and lecture room. Brighton did not impress him, perhaps because (to a severely disciplined mind) it owed its existence to the pursuit of pleasure. In a letter to his younger sister Margaret, only the Chain Pier, combining use with ornament, receives any approbation:

> I do not at all admire Brighton, i.e. its character as a fashionable or interesting place. It is very convenient for distance, lodging, accommodation, food, &c.—but these are not what I refer to; I mean as to its beauties, natural or artificial, or as to its importance in advancing great interests, as civilisation or improvement.

> Considered in this way, Brighton is to me very commonplace and poor: there are no natural beauties there to distinguish

it from a thousand other places; there are no high interests concerned to raise it above the poor distinction of being a place resorted to by company because other company was there before them. As to the Pavilion, there is scarcely a single cottage in or about the poor village of Crab Niton that does not both in beauty and use surpass it. The Pavilion has no beauty for the painter; and what is intended for beauty, of which there is a great deal, has no use.

The Steine is a good street, and many of the squares and places are good, also many of the old houses; and could one but see a sufficient cause why they had come together—i.e. the presence of any beautiful or useful feature—the town, with the exception of one or two things, would be a very good one. It has, however, one thing perfectly beautiful every way in the chain-pier—an admirable specimen of ingenuity and art, and which, destined to useful purposes, not only pleases the eye but satisfies the mind.

Nevertheless, delight in nature and its ephemeral phenomena is never far from the surface, as appears from the way he uses weather for a metaphor to encourage his brother-in-law Edward. Implicit, too, is an approach to the understanding of nature that, while becoming more precise as error is eliminated, stops short before that complete knowledge only to be attained (as Faraday's religion would tell him) in the hereafter. One almost expects to read the words 'through a glass, darkly':

Mr dear Edward,—I intended to have written you a letter immediately upon the receipt of yours, but delayed it, and perhaps shall not now say what occurred to me then. Why do you write so dully? Your cogitations, your poetry, and everything about your letter, except the thirty pounds, has a melancholy feel. Perhaps things you had scarcely anticipated are gathering about you, and may a little influence your spirits; and I shall think it is so for the present, and trust it is of but little importance, for I can hardly imagine it possible that you are taken unawares in the general picture of life which you have represented to yourself: your natural reflection and good sense would teach you that life must be chequered, long before you would have occasion to experience it. However, I shall hope this will find you in good spirits and laughing at such thoughts as those in which you were immersed when you wrote me.

I have been watching the clouds on these hills for many evenings back: they gather when I do not expect them; they throw down rain to my mere inconvenience, but doing good to all around; and they break up and present me with delightful and refreshing views when I expect only a dull walk. However strong and certain the appearances are to me, if I venture an internal judgement, I am always wrong in something; and the only conclusion I can come to is, that the end is as beneficial as the means of its attainment are beautiful. So it is in life; and though I pretend not to have been much involved in the fogs, mists, and clouds of misfortune, yet I have seen enough to know that many things usually designated as troubles are merely so from our own particular view of them, or else ultimately resolve themselves into blessings.

Do not imagine that I cannot feel for the distresses of others, or that I am entirely ignorant of those which seem to threaten friends for whom both you and I are much concerned. I do feel for those who are oppressed either by real or imaginary evils, and I know the one to be as heavy as the other. But I think I derive a certain degree of steadiness and placidity amongst such feelings by a point of mental conviction, for which I take no credit as a piece of knowledge or philosophy, and which has often been blamed as mere apathy. Whether apathy or not, it leaves the mind ready and willing to do all that can be useful, whilst it relieves it a little from the distress dependent upon viewing things in their worst state.

The point is this: in all kinds of knowledge I perceive that my views are insufficient, and my judgement imperfect. In experiments I come to conclusions which, if partly right, are sure to be in part wrong; if I correct by other experiments, I advance a step, my old error is in part diminished, but is always left with a tinge of humanity, evidenced by its imperfection. The same happens in judging of the motives of others; though in favourable cases I may see a good deal, I never see the whole. In affairs of life 'tis the same thing; my views of a thing at a distance and close at hand never correspond, and the way out of a trouble which I desire is never that which really opens before me.

Now, when in all these, and in all kinds of knowledge and experience, the course is still the same, ever imperfect to us, but terminating in good, and when all events are evidently at the disposal of a Power which is conferring benefits continually upon us, which, though given by means and in ways we do not comprehend, may always well claim our acknowledgement at last, may we not be induced to suspend our dull spirits and thoughts when things look cloudy, and, providing as well as we can against the shower, actually *cheer our spirits* by the thoughts of the good things it will bring with it? And will not the experience of our past lives convince us that in doing this we are far more likely to be right than wrong?

Your third page I can hardly understand. You quote Shakespeare: the quotation may be answered a thousand times over from a book just as full of poetry, which you may find on your shelf. The uses of the world can never be unprofitable to a reflecting mind, even without the book I refer to; and I am sure can only appear so to you for a few hours together. But enough of this; only, when I get home again, I must have a talk with you. The money and our friends came safe. Give our love to Caroline and the rest of our family.

Michael and Sarah had no children of their own, but delighted the company of numerous nephews and nieces 'Unkle' Michael

Michael and Sarah Faraday.

never let his interest in natural phenomena lapse, especially when replying to a report from Andrew Reid, the son of Sarah's sister, who had described what was clearly a narrow escape from being struck by lightning:

> Your aunt shewed me your last letter which, besides the pleasure it gave as communication, I read with great interest on account of its philosophical character and so, to divide things properly, we are about to return you a joint answer in which all the useful will be touched upon by your Aunt & the fantastical (as you may call it) by me.
>
> I envy you your stormy evening walk, and think that if we had been together we might have made some very interesting observations. I do not doubt any part of your description, & think that the hissing and luminous appearance which you saw were the natural preparations and precursors of

the discharge of lightning which immediately afterwards occurred.   Did you happen to observe any luminous appearance on yourself? I have little doubt that if you had held out your hand, each finger would have had the light on the tip. Perhaps, however, it is as well that you did not happen to do so. It seems to me very likely that the lamp post was struck by the lightning, or at all events something near it, i.e. in the neighbourhood.

Did you see the light go out? Have you seen the post since, & examined the bars to see whether they seemed affected? Is the whole post iron or part iron, part wood? Is it a gas lamp or an oil lamp post? The effect of the snow in making the post brighter is very curious. How was the night altogether? Very dark? Or was there a sort of luminous haze about?

I have sent you a lot of questions, but you will perceive they are so many proofs that I believe your story. And now I may say, I do not think your situation was without danger and am very happy you escaped, to be a comfort to your parents & family, to whom give the affectionate love of your affectionate Unkle.

In similar vein, to a great niece he describes an electric eel (subject of experiments much earlier in his career) and slips in a comment about experimenting with a magnet. Was Constance's curiosity sustained by talks with her uncle?

First a kiss—s—s—s—ss. Next, thank you for your good letter—very well written and very pleasant—and now thanks for the letter you are going to write me in which you must tell me how Papa & Mama do—and what you are about. I went this morning to see a fish like a great eel take his breakfast. This morning he had three frogs for breakfast—yesterday he eat [sic] 9 fish in the course of the day, each as large as a sprat, and the day before, 14. When the fish are put into the water he electrifies and kills them & then swallows them up. And if a man happens to have his hands in the water at the same time the fish (that is, the eel) electrifies the man too. The eel is now above 12 years old and is heavier, I think, than you are.

Yesterday I saw the Royal children, the Prince of Wales and the Duke of York. Such nice children, they would make famous playmates for you but I do not know whether Princes do play much. I do not think they can be so happy in their play as you are.

As to the magnet, when you & I meet we will have a long talk about it and make some *experiments*—

and so with my love to Papa & Mamma and curious Constance with a kiss for each I am.

> Your loving old Uncle
>
> M. Faraday

The Sandemanian Church, that small nonconformist group in which he was brought up, played a major role in Faraday's life, though it does not figure greatly in his writings. An exception to this silence occurs when he was appointed an elder of the church in 1840, a role which clearly filled him with awe, and a large sense of inadequacy (certainly more than he would ever have evinced in the face of the most eminent of his scientific peers). Writing to Edward Buchanan, an elder of the church in Edinburgh, scriptural quotation provides both warning and comfort in the task he had been asked to fulfil:

I am very grateful for your comforting letter and cannot say how much I value it. It came to me as a warming of love, & the strengthener of hope at a time when my thoughts were very foolish, and I hope I shall never cease to prize & profit by it. You know the working of the heart pretty well, but you can hardly imagine my folly and the distress of mind which came over me at the Church's call: I doubted not that it was of the Lord's doing, but my utter unworthiness & unfitness made me tremble with the fear of its being to the increase of condemnation at that day when God would judge all men by Jesus Christ, for I thought of what he says to Pharaoh for this same cause: 'have I raised thee up that I might shew forth my power in thee'. But the law of the brethren & their consolations (and above all, I hope, the power of God) has removed much of this trouble, though I cannot but fear & tremble; and then again when I feel comforted and happy, my foolish heart suggests that I am perhaps relieved only by feeling slightly the vast importance of the character which I have ventured to accept.

But I endeavour to cast off such thoughts, & comfort myself by the many promises that He will be the strength of His people; that He will give them a mouth & wisdom which all their adversaries cannot gainsay; that His strength is made perfect in weakness; that it is not any thing that we have done or can do, but for His own name's sake He works making His people willing in the day of His power; and surely there is enough (and far more than enough) to make even such a foolish heart as mine rejoice & have confidence in the Lord.

Chapter 3

Touring the Continent: 1813–1815

Only a few months after he joined the Royal Institution as Humphry Davy's 'chemical assistant', young Michael Faraday was presented with a wonderful and unlooked for opportunity that was to add immeasurably to his education and bring him into contact with some of the foremost scientists in Europe. The year before, Davy had married a rich widow, Mrs Apreece, and he determined on an extended tour on the continent, combining culture and social amusements with visits to scientific colleagues. At this time Britain was still at war with France, so at first sight such a project must have seemed audacious, not to say foolhardy. However, several years previously, the Institut de France had awarded Davy a prize created by Napoleon (who had always taken a personal interest in scientific affairs) for the most meritorious work on electricity. Hence Davy's application for visas was granted, though one might be forgiven for believing (with some of the British press) that a strong element of somewhat subversive public relations might have influenced the French government's invitation. Be that as it may, on 13 October 1813 the party set out. At first the plan had been for the Davys to be accompanied by Sir Humphry's valet, La Fontaine who was originally from Flanders, but his wife would not allow it, in the belief that should he set foot again in France he might never return to London. So it came about that Davy asked his young assistant to join them instead, and carry out a few extra menial duties till they reached Paris, where they would hire a valet. For Faraday, who had never been more than a few miles from home in his life, the chance to undertake the kind of 'Grand Tour' favoured by the young aristocracy was an astonishing piece of good fortune. From the first to the last day of the journey he kept a diary, whose freshness and spontaneity still shine from the page; he was just 22 years old:

Wednesday 13th. This morning formed a new epoch in my life. I have never before, within my recollection, left London at a greater distance than twelve miles and now I leave it perhaps for many years, and to visit spots between which and home whole realms will intervene. 'Tis indeed a strange venture at this time to trust ourselves in a foreign and hostile country, where also so little regard is had to protestations and honour that the slightest suspicion would be sufficient to separate us for ever from England and perhaps from life. But curiosity has frequently incurred dangers as great as these and therefore why should I wonder at it in the present instance? If we return safe, the pleasure of recollection will be highly enhanced by the dangers encountered, and a never failing consolation is that, whatever be the fate of our party, variety (a great source of amusement and pleasure) must occur.

Friday 15th. Reached Plymouth this afternoon. I was more taken by the scenery today than by anything else I have ever seen. It came upon me unexpectedly and caused a kind of revolution in my ideas respecting the nature of the earth's surface. That such a revolution was necessary is, I confess, not much to my credit and yet I can assign to myself a very satisfactory reason in the habit of ideas induced by an acquaintance with no other green surface than that within 3 miles of London. Devonshire, however, presented scenery very different to this. The mountainous nature of the country continually put forward new forms & objects and the landscape changed before the eye more rapidly than the organ could observe it. This day gave me some ideas of the pleasures of travelling and have raised my expectations of future enjoyments to a very high point.

From Plymouth the party sailed to Morlaix. Young Michael's first sea voyage gave him the opportunity to observe the phosphorescence of the waves at night and to try out his resistance to sea sickness on the rough sea:

Monday 18th. I last night had a fine opportunity of observing the luminous appearance of the sea and was amused by it for a long time. As the prow of the vessel met the waters it seemed to turn up a vast number of luminous bodies about the size of peas, some, however, being larger than others. These appeared to roll onwards by the side of the vessel with the waters and sometimes traversed a distance of many yards before they disappeared. They were luminous at, or beneath, the surface of the water indifferently, and the only effect produced by different depths was a diminution of the light by the quantity of intervening medium. These luminous spots were very numerous, the most (so I think) about half an hour after midnight. Their light was very bright and clear.

The swell on the sea was very considerable all night, though gradually decreasing. I remained on deck and escaped all sea sickness. As day came on, and the light increased, we looked about us but saw nothing in the scene except sky and immense waves striding one after the other at a considerable distance. These, as they came to us, lifted up our small vessel and gave us (when on their summits) a very extended horizon, but we soon sank down into the valleys between them and had nothing in view but the wall of waters around us.

While the travel and customs arrangements were being made at Morlaix, Michael absorbed impressions of his first French town: what he saw did not please him, though he was inclined to give the benefit of the doubt to French cooking:

Thursday 21st. I have been wandering into the town today, for, though there is sufficient in the house to surprise me at every turn, yet I wished to see the general appearance of things. I find them, however, all alike and I cannot refrain from calling this place the dirtiest and filthiest imaginable. The streets are paved from house to house with small sharp stones, no particular part being appropriate to foot passengers. The kennels are full of filth and generally close to the house. The places and corners are occupied by idle loiterers who, clothed in dirt, stand doing nothing and the houses—but I will endeavour to describe our hotel. This (the best in the place) has but one entrance, and it is paved in a manner similar to the streets; through it pass indiscriminately horses, pigs, poultry, human beings or whatever else has a connection with the house or the stables and pigsties behind it. On the right hand of the passage, and equally public as a thoroughfare with it, is the kitchen. Here a fire of wood is generally surrounded by idlers, beggars or nondescripts of the town who meet to warm themselves and chatter to the mistress, and they hold their stations most tenaciously though the processes of cooking are in progress. I think it is impossible for an English person to eat the things that come out of this place except through ignorance, or actual and oppressive hunger, and yet perhaps appearances may be worse than the reality, for in some cases their dishes are to the taste excellent and inviting, but then they require, whilst on the table, a dismissal of all thoughts respecting the cookery or kitchen.

The benefits of a good British cup of tea soon made themselves apparent, as did the attractions of picnicking in France:

Tuesday 26th. —We left Laval before breakfast this morning, and were afterwards sorry we had done so. About 11 o'clock that meal became necessary, but we could meet with no accommodations to get it. We now found the utility of English contrivance and resorted with success to a travelling tea-kettle and caddy. Contributing the tea, sugar and kettle on our part, we were able to heat some water at a Post house at which we stopped and, making milk from the yolk of eggs, produced the necessary beverage. Bread was deficient and, not liking that of the house, we sought for some at a baker's shop, where it was found of an excellent quality. Thus luckily our fears of fasting were removed and we were reinstated in a travelling condition.

Thursday 28th. —I cannot help dashing a note of admiration to one thing found in this part of the country—the Pigs! At first I was positively doubtful of their nature, for though they have pointed noses, long ears, rope like tails and cloven feet, yet who would have imagined that an animal with a long thin body back and belly arched upwards, lank sides, long slender feet and capable of outrunning our horses for a mile or two together, could be at all allied to the fat sow of England?

Arriving in Paris, Michael found further ammunition for waspish disapproval of the way things were ordered in France, though he was ready to make an effort by learning the language:

Friday 29th. —I am here in the most unlucky and irritating circumstances possible. Set down in the heart of Paris, that spot so long, so desiringly looked at (& so vainly, too, from a distance) by numbers of my countrymen. I know nothing of the language, or of a single being here, added to which the people are enemies & they are vain. My only mode will be to stalk about the town looking (and looked at) like a man in the monkish catacombs. My mummies move, however, and they see with their eyes: I must exert myself to attain their language so as to join in their world.

The streets of Paris are paved with equality, that is to say, no difference is made in them between men & beasts, and no part of the street is appropriated to either. Add to this that the stones of which the pavement consists are very small and sharp to the foot, and I think much more need not be said in praise of it.

S. 14th. —I went this morning into some of the churches but was not induced to stop long in any of them. I could hardly have expected that they would have attractions for a tasteless heretic. There were many people in some of them but numbers seemed like me to be gazers. A theatrical air spread through the whole and I found it impossible to attach a serious or important feeling to what was going on.

Faraday soon made his first visit to a French laboratory. The compound which maimed its discoverer M Dulong was nitrogen trichloride. Michael's predicament over his passport recalls the worries of many young travellers before and since:

T. 18th. I had occasion to go this morning to M Vauquelin. M Dulong, the discoverer of the new detonating compound of chlorine and azote, is engaged in this laboratory. He was

The French chemist M Vauquelin.

not there at this time but the person in the warehouse spoke of him. He lost an eye by an explosion of the compound at the time that he first discovered it. I found last night that I had lost my passport. I do not know what the police will say to me about it.

Further trenchant comparisons of British and French ways are still largely to the disadvantage of the latter:

W. 24th. Being indoors all day, I amused myself by noticing in what the apartments we occupy differ from English rooms. The most striking difference in this cold weather is in the fuel and fire places. Wood is the universal fuel, at least from Morlaix to this place, and I understand there are very few places where they burn coals. If you keep up a cheerful fire, it is very expensive and if you keep up a dull fire, which is done by smothering up the wood with the ashes, it is but a sorry sight. I can never comfortably make a comparison between them and an English coal fire. To conclude, French apartments are magnificent, English apartments are comfortable; French apartments are highly ornamental, English apartments are clean; French apartments are to be seen, English apartments enjoyed, and the style of each kind best suits the people of the respective countries.

Books are very cheap here in proportion to English books. I should think on an average they are scarcely half the price and yet large private libraries are seldom met with. Bibliomania is a disease apparently not known in France; indeed, it is difficult to conceive how their light airy spirits could be subjected to it.

Humphry Davy carried with him a travelling kit for carrying out simple experiments in chemical analysis. Consequently, when a strange new substance, deep violet in colour, was drawn to his attention, he immediately went to work on it. It had been extracted for the first time from seaweed two years earlier by Monsieur

Courtois, though at the time Davy was shown it, the method had not been published. The substance was iodine. To Faraday, this unexpected discovery was a symbol of rich intellectual rewards to be had from experimental chemistry:

> W. Dec. 1. On this and the preceding days Sir H. Davy made many new experiments on the substance discovered by M. Courtois. The discovery of this substance in matters so common, and supposed so well known, must be a stimulus of no small force to the enquiring minds of modern chemists. It is a proof of the imperfect state of the science, even in those parts considered as completely understood. It is an earnest of the plentiful reward that awaits the industrious cultivator of this, the most extensive branch of experimental knowledge. It adds in an eminent degree to the beautiful facts that abound in it, and presents another wider field for the exercise of the mind. Every chemist will receive it as an addition of no small magnitude to his knowledge, and as the forerunner of a grand advance in chemistry.

Bargaining for purchases was clearly a novelty for the Englishman:

> T. 21. I am quite out of patience with the infamous exhorbitance of these Parisians. They seem to have neither sense of honesty nor shame in their dealings. They will ask you twice the value of a thing with as much coolness as if they were going to give it you, and when you have offered them half their demand (and on their accepting it), you reproach them with unfair dealings, they tell you 'you can afford to pay'. It would seem that every tradesman here is a rogue, unless they have different meanings for words to what we have.

From Paris the party travelled south, following the main road through Fontainebleu, Nemours and Moulins that is now the Route Nationale 7. The beauty of his surroundings, added to thoughts about the uniqueness of his situation, brighten the tone of the narrative, much as travelling southwards along the N7 has raised the spirits of so many of Faraday's fellow countrymen before and since. Even the hotels and the food began to give more pleasure:

W. 29. This morning we left Paris after a residence in it of three months and prepared ourselves for new objects and new scenes. The morning was fine, but very cold and frosty; but on entering the forest of Fontainebleu we did not regret the severity of the weather, for I do not think I ever saw a more beautiful scene than that presented to us on the road. A thick mist which had fallen during the night, and which had scarcely cleared away had, by being frozen, dressed every visible object in a garment of wonderful airiness & delicacy. Every small twig and every blade of herbage was encrusted by a splendid coat of hoar frost, the crystals of which in most cases extended above half an inch. This circumstance, instead of causing a sameness (as might have been expected), produced an endless variety of shades and forms. Openings in the foreground placed far-removed objects in view, which, in their airiness (and softened by distance) appeared as clouds fixed by the hands of an enchanter: then rocks, hills, valleys, streams and road; then a mile stone, a cottage or human beings came into the moving landscape and rendered it ever new and delightful.

T. 30. Though cold and dark, we were on our way to Moulins by 5 o clock this morning and, though somewhat more south than London, yet I do not perceive any superior character in the winter mornings here. However, as we always judge worse of a bad thing when it is present than at any other time, I may have been too cross with the cold and dark character of our early hours. The moon had set a circumstance to be regretted for, though assisted only by the faintness of star light, yet I am sure our road was beautiful. 'Twas along the bank of the river and within a few yards of the water, which indeed at times came to the horses feet. On our left was a series of small hills and valleys, lightly wooded and varied now and then by clustering habitations. These dark hours, however, have their pleasures, and those are not slight which are furnished at such hours by the memory or the imagination.

'Tis pleasant to state, almost audibly to the mind, the novelty of present circumstances: that the Loire is on my right hand; that the houses to the left contain men of another country to myself; that it is French ground I am passing over, and then of the distance between myself and those who alone feel an interest for me, and to enjoy the feeling of independence and superiority we at present possess over those sleeping around us. We seem tied to no spot, confined by no circumstances, at all hours, at all seasons and in all places. We move with freedom; our world appears extending and our existence enlarged. We seem to fly over the globe rather like satellites to it than parts of it, and mentally take possession of every spot we go over.

The Hotel we are in is a very excellent one and it is but fair that, whilst marking down the singular wants of places professing to be for accommodation, I notice also such as answer to their professions. The house is large, warm and actually comfortable. It perhaps might not have this last point of character in England because of a little dirt, a little too much of publicity in the different parts of it and some little privations, but standing, as it does in a French town, it is eminently so. The accommodations (particularly with respect to food) are excellent; good victuals, good wine, good fruit and plenty of all are great things to travellers, and when to these are added a tolerably clean bed and bedroom, great inducements to halt a day or two are presented. Lastly, you always have a cheerful obliging host. He will answer you any question; he says he will do any thing for you, and frequently does do things apparently entirely from good nature: and though a suspicion may lurk in the mind that these things are intended to make you content with the bill when you get it, yet even the appearance of liberal good nature is pleasant and *will* have its effect.

S. 15. Bread, wine and water are three things very good and very cheap here. Of bread, there are many varieties. What they call *pain d'ammunition* looks more like brick than bread; the price is almost nothing, and for that reason much of it is eaten (for the poor people are very poor here). The better sorts of bread are very good and the best is more like cake than bread. I think the character the French bread has of surpassing that of most other nations must be well founded. It has a degree of positive excellence that one would not wish it surpassed; beyond what it is would be undue luxury.

Arriving in the Midi, they came first to Montpellier, where nocturnal walks showed the city without the inconvenience of bumping into its inhabitants:

I have felt much pleasure during a few night strolls about the town in viewing the fountains, but the pleasure was heightened by other circumstances. Although frosty, very little snow has yet fallen, so that the streets are very clean and dry. The nights are generally clear and serene. The moon has shone bright and the inhabitants have been housed. In this state I find a town very interesting. You see the latest possible marks of human beings; you see everything constructed for their use and convenience, but you see none of the persons themselves, and may easily imagine them far away. It is, indeed, the best possible situation for the mind to abstract man from the artificial supplies which he obtains for his wants equally artificial, and to consider them independent of each other.

Next came Provence, to be followed by a crossing of the Alps via the Col de Tende. It was now February 1814 and the Maritime Alps were covered in snow. While Lady Davy was transported in a kind of sedan chair, after the road gave out at Tende, Michael (suitably muffled against the cold) and Sir Humphry (who took the opportunity of pointing out the sites of geological interest) walked:

Montpellier (top); Avignon (bottom) in 1820.

F. 11. Today took a walk round Aix and the environs. The weather is warm and beautiful, vegetation is fast advancing, everything seems cheerful and summer on the verge of

View of Nice in 1820.

coming. I found many small green lizards basking on a bank of lettuces in the rays of the sun. They were too nimble to be caught.

T. 17. Left Nice this morning and advanced towards the Alps by a road on the sides of which were gardens with oranges and lemons in great profusion. We soon entered among the mountains; they were of limestone, stratified very regularly and appearing at a distance like stairs. At some distance up, we came to a place where the strata for many yards consisted of small pieces of limestone of an inch in size, more or less cemented together by carbonate of lime. Varieties occurred here and there and in these places the cement had taken a stalactitic form. A dropping well added to the variety of objects and, appearing in a very picturesque situation, added much to the beauty of the scene.

On the steep sides of the mountains, men and women were gathering olives. Here the olive trees assume a much larger size and better appearance than in the South western parts of France and add much to the beauty of the foreground in a mountainous landscape, for, though the verdure is pale and not to be compared with that of the pine, yet the trunk has great variety of form and much beauty. To collect the fruit, which is now ripe black and of a very bitter taste, they spread large sheets of cloth under the trees and then men ascend and knock off the fruit with poles. A slight concussion or shaking is sufficient for this purpose, for the ripe fruit is easily separated from the branches.

We had now got considerably up the mountain and the valley behind us appeared like a map. By the road side was a wild fig tree still bearing last year's fruit. On the opposite mountains a flock of sheep was grazing, but they seemed more like maggots than what they were. The sea was soon visible over the tops of the lower mountains, and snow began to surround us. Having arrived at the top, we looked back and were amply repaid for our pains by the beauty and singularity of the scene. There was the deep valley we had left, spotted with white houses, & the little bridges of the river that ran through it. The sheep halfway down the mountain side, and the winding road, enlivened the tremendous precipices and the sea in the distance splendidly reflected the rays of a bright sun. At Tende the noble road which had given such facile and ready conveyance finished, & it was necessary to prepare for another sort of travelling. Expecting it would be very cold, I added to my ordinary clothing an extra waistcoat, two pair of stockings and a night cap. These, with a pair of very strong thick shoes and leathern overalls, I supposed would be sufficient to keep me warm.

On each side were extensive plains covered with snow to a great depth, but sufficiently hard and solid to support the men who accompanied, or rather who fetched us, as they walked upon it. There were at present but two of these persons, the chief and one of the sixty five composing the band. They walked on before, whistling, and the scene (so strange and singular to us) never attracted their attention, unless to point out to us the site of an avalanche or a dangerous place.

There was something pleasant in the face and appearance of the chief and I thought him a good specimen of the people here. He was a tall man, not at all thick, but his flesh seemed all muscle and strength. His dress consisted of few articles: trousers, a loose waistcoat, an open jacket, a hairy cap, very heavy soled shoes and coarse gaiters or overalls tied round his shoes to keep out the snow was all his clothing, and I found his comrades just like him. His gait was very peculiar, contracted (I suppose) by walking constantly on the snow, where a firm footing is required.

The road began to change soon after leaving Tende, and at last became nothing but ice. It was fit for beasts of burden only: grooves had been formed in it at equal distances to receive the feet of the horses or mules and prevent their falling, and though convenient to them, it was to us a great evil, for as the wheels fell successively into the ruts, it produced a motion not only disagreeable but very dangerous to the carriage. Sir H Davy here pointed out to me the rocks of micacious shist and I learned at the same time that granite is always found under this rock. The only vegetation visible (though there might be much under the snow) was of pine trees; they lifted their verdant tops above the snow and in many places broke the monotony of a white landscape.

At 1 o'clock we began to ascend the mountain and I commenced walking with a barometer in my hand. The path quickly changed its appearance and soon became not more than 18 or 24 inches wide. Being formed by the constant tread of mules, it consisted merely of a series of alternate holes in the snow. After some climbing and scrambling (the exertion of which was sufficient to keep me very comfortably warm), I reached a ruined desolate house half way up the mountain.

After a short rest all resumed their labour and at 43' after 3 o'clock I gained the summit of the mountain, having been

3 hours ascending.   Here at a height of more than 6000 feet above the level of the sea, the thermometer was at 11 °F and the barometer at 25.3 inches.   The summit of the mountain is very pointed and the descent consequently begins immediately on the other side. To descend was a task which, though not so tedious, was more dangerous than to ascend. The snow was in much greater quantity on this side of the mountain than on the other and in many places where it had drifted, assumed a beautifully delicate appearance. In numberless spots it was (according to the men) more than 20 feet deep and in descending, it often received me more than half way into it.

Finally, on the Italian side of the mountain, they could continue by carriage (albeit through deep snow), and at length reached Turin, where it was carnival time.   Such a long and detailed description of the happenings confirms young Michael's keen (if detached) interest, combined with a reporter's talent for description:

Leaving Vernon we proceeded in the same way through roads (or rather paths) so bad and narrow that it was a matter of surprise how the carriage escaped with only a large scratch, for the walls of snow here were almost as formidable as walls of stone. After a time, appearances began to improve and signs of cultivation were again visible in the banked up sides of the mountains, and small birds were flying about; these were to us very cheerful appearances.   The snow, instead of being hard and dusty, was soft and adhesive and the river which had commenced in the mountains and often accompanied us, was now very large. We met many small sledges drawn by mules.

This day happened to be the last of the Carnivale. So I walked out in the afternoon to see what was doing in public. Towards three o clock, shops shut up very rapidly and the masters betook themselves to walking, gazing and the amusements now going on. Such as were determined to be cheerful, in spite of appearances, joined the numbers that were waltzing to the music of itinerant musicians, and certainly these did not seem the least cheerful and happy part of the populace of Turin (I may perhaps add also were not the least numerous). I strolled to one place just on the skirts of the town & found it crowded by those who thus easily obtained the pleasures. It was a large clear piece of ground on the bank of a branch of the Po and resounded from end to end, side to side, with the harmony of a number of musical professors. The little groups into which they had formed themselves were surrounded each one by its circle of ever-moving and never-tired dancers, and spaces between these groups were filled up by a heterogeneous mixture of singers, leapers, boxers, chestnut merchants, apple-stalls, beggars, trees and lookers on. I fell in with one of the most worthy sets (at least they claimed the pre-eminency, and it was allowed them by the other mobs). The nucleus was an enormous stone on which stood totteringly five musicians and twenty one pair were waltzing round them.

Returning into the town, I found that those of Turin, who were superior to the vulgar amusements I have just described, had resorted to the employment which custom has ascertained to be more refined and suited to their ordinary habits and occupations. That such a suitableness exists, I verily believe, but I think I perceived much more cheerfulness (and means much better suited to produce it) in the crowds I had left than in those I came to see; but pride will supply man wants, and food, clothing, amusement and comfort are very often given up for its peculiar gratification. I found myself in a wide and spacious street of considerable length, terminating at one end in a large place having a church in its centre. All the entrances into this street were guarded by soldiers, and no person on horseback or in a carriage could gain admission into it, except at the top of a long string of carriages. Curricles, saddle horses &c filled it, and they continued to move on progressively up and down the street

and round the church for several hours. It was presumed
that these vehicles carried the principal persons of the town
but nobody pretended to say that the owners were actually
in them. One of very goodly aspect and fine appearance
was pointed out to me when coming up. The horses were
very handsome, and the coachman and footman as spruce
as could be, and somewhere the two maids in the inside.
The next was not so dashing, but it was empty and the
third was so shabby that I did not look to see what was
inside.

There were, however, an immense number of persons who
stood on each side of the street, looking and gazing with
great apparent satisfaction and who, if they had been
conscious of the comparison I was then making between
the scene before me and the one I had just left, would
have looked down upon me with contempt & derision, no
doubt equal at least to that which, at the same moment,
occupied my mind. Silly, however, as the whole affair
was, it had nearly led to circumstances of more importance.
A gentleman in his curricle, attended by his servant, had
come down one of the side streets and wished to enter
the 'Corso' unlawfully but was opposed by the soldier
guarding the entrance. The gentleman, irritated by the
repulse, endeavoured to force his way by rough driving. The
soldier set his bayonet and stood his ground. The horse was
slightly wounded and near being killed, and from the pain
became restive and had nearly killed his master, who was
in the end obliged to turn back with his wounded horse
amidst the derision and laughter of the surrounding mob.
As evening came on, the Corso gradually broke up and the
Carnivale concluded with numberless theatres and a masked
ball.

From Turin they moved on to Genoa and continued south
along the coast of Liguria, where Faraday recounts a most unusual
meteorological event: not just one, but three water spouts out

The Gulf of La Spezia in 1820.

to sea. His description is characteristically precise but he is still careful in stressing it was subjective:

> The weather as yet against our voyage and in the afternoon a storm of thunder, lightning & rain with waterspouts. A flash of lightning illuminated the room in which I was reading, and I then went out on the terrace to observe the weather. Looking towards the sea, I saw three waterspouts all depending from the same stratum of clouds. I ran to the sea shore on the outside of the harbour, hoping they would approach nearer, but that did not happen. A large and heavy stratum of dark clouds was advancing, apparently across the field of view, in a westerly direction. From the stratum hung three water spouts, one considerably to the west of me, another nearly before me and the third eastward. They were apparently at nearly equal distances from each other. The one to the west was rapidly dissolving, and in the same direction a very heavy shower of rain was falling, but whether in the same place, or nearer or more distant, I could

Sketch of a waterspout from Faraday's journal.

not tell: rain fell violently all the time at Genoa. The one before me was more perfect and distinct in its appearance. It consisted of an extended portion of cloud, very long and narrow, which projected from the mass above downward in a slightly curved direction towards the sea: this part of the cloud was well defined, having sharp edges, and at the lower part tapering to a point. It varied its direction considerably during the time that I observed it, sometimes becoming more inclined to the horizon and sometimes less; sometimes more curved and at others more direct. Beneath this projecting cloud, and in a direction opposite to the point, the sea appeared violently agitated. At the distance they were from me, I could merely perceive a vast body of vapour rising in clouds from the water and ascending to some height, but disappearing (as steam would do) long before it reached the point of the cloud.

The destruction and dissolution of the water spouts seemed to proceed very rapidly when it had once commenced and 3 or 4 minutes after the apparent commencement of decay it had entirely disappeared. The vapour, the sea and the cloud diminished in nearly equal proportions. They were situated much farther out to sea than I at first supposed (I should think 5 or 6 miles) and of course, what I have here noted is merely a relation of the thing as it appeared to me and is possibly very different to the real truth.

Towards the end of March 1814 the party arrived in Florence where, after a few days sightseeing, preparations were made to carry out a unique experiment. Since ancient times there had been speculation about the constitution of diamond: was it merely a form of carbon, like graphite, or did it contain other elements? How it could be that a chemical element might exist in more than one form (especially where they were so different as diamond and graphite) was not at all clear. A direct means to discover whether the diamond contained anything other than carbon would be to burn it and analyse any residue and products. Now of course, diamonds do not burn easily, and extremely high temperatures are required, but it happened that the Grand Duke of Tuscany had in his possession a very large burning glass, consisting of two lenses more than a foot in diameter by means of which the sun's rays could be focused. The experiment was a success and the result unequivocal: diamond was nothing but a form of carbon:

> Today we made the grand experiment of burning the diamond, and certainly the phenomena presented were extremely beautiful and interesting. The Duke's burning glass was the instrument used to apply heat to the diamond. It consists of two double convex lenses. The instrument was placed in an upper room of the museum and, having arranged it at the window, the diamond was placed in the focus and anxiously watched. The heat was thus continued at intervals for 3/4 of an hour (it being necessary to cool the globe at times) and during that time it was thought that the diamond was slowly diminishing and becoming opaque. On a sudden, Sir H Davy observed the diamond to burn visibly, and when removed from the focus it was found to

be in a state of active and rapid combustion. The diamond glowed brilliantly with a scarlet light, inclining to purple and, when placed in the dark, continued to burn for about four minutes. After cooling the glass, heat was again applied to the diamond and it burned again, though not nearly so long as before. This was repeated twice more and soon after, the diamond became all consumed. This phenomenon of actual and vivid combustion, which has never been observed before, was attributed by Sir H Davy to the free access of air; it became more dull as carbonic acid gas formed and did not last so long. From these experiments, according to Sir H Davy, it is probable that diamond is pure carbon.

Ten days later, they came to Rome where, among his visits to the sights, Faraday found his old trade of bookbinder being carried on, though he was not impressed by the results:

T. 7. Walked about the City to gain a general and local knowledge of it. Went into a bookseller's shop to enquire for an Italian and English dictionary, but could not find one. Went into the workshop of a bookbinder and saw there the upper part of a fine corinthian pillar of white marble, which he had transformed into a beating stone of great beauty. Found my former profession carried on here with very little skill, neither strength nor elegance being attained.

From time to time Davy dealt with his own correspondence by entrusting it to a courier, giving Faraday the opportunity of enclosing any missives of his own in the package to be conveyed to England. Thus, on 14 April 1814, Faraday seized the chance to address a long letter to his mother. Much of it is filled by straightforward descriptions of what he had seen, but there are also some affecting personal passages:

You must consider this letter as a kind of general one, addressed to that knot of friends who are twined round my heart; and I trust that you will let them all know that, though distant, I do not forget them, and that it is not from want of regard that I do not write to each singly, but from want of convenience and propriety; indeed it appears to me that there is more danger of my being forgot than of my forgetting. The first and last thing in my mind is England, home and friends. It is the point to which my thoughts still ultimately tend and the goal to which, looking over intermediate things, my eyes are still directed. But on the contrary in London you are all together, your circle being little or nothing diminished by my absence; the small void which was formed on my departure would soon be worn out and, pleased and happy with one another, you will seldom think of me. Such are sometimes my thoughts, but such do not rest with me; an innate feeling tells me that I shall not be forgot, and that I still possess the hearts and love of my mother, my brother, my sisters and my friends.

When Sir H. Davy first had the goodness to ask me whether I would go with him, I mentally said, 'No, I have a mother, I have relations here.' And I almost wished that I had been insulated and alone in London, but now I am glad that I have left some behind me on whom I can think, and whose actions and occupations I can picture in my mind. Whenever a vacant hour occurs, I employ it by thinking on those at home. Whenever present circumstances are disagreeable, I amuse myself by thinking on those at home. In short when sick, when cold, when tired, the thoughts of those at home are warm and refreshing balm to my heart. Let those who think such thoughts useless, vain and paltry think so still; I envy them not their more refined and more estranged feelings. Let them look about the world unencumbered by such ties and heart-strings and let them laugh at those who, guided more by nature, cherish such feelings. For me, I still will cherish them in opposition to the dictates of modern refinement as the first and greatest sweetness in the life of man.

A letter to his friend Benjamin Abbott also conveys the impression of vivid sensation that the tour had brought to Faraday:

It is now 9 months ago since I left London, but I have not forgot (and never shall forget) the ideas that were forced on my mind in the first days. To me, who had lived all my days of remembrance in London, in a city surrounded by a flat green country, a hill was a mountain and a stone, a rock; for though I had abstract ideas of the things, and could say 'rock' and 'mountain', and could talk of them, yet I had no perfect ideas. Conceive, then, the astonishment, the pleasure and the information that entered my mind in the varied county of Devonshire, where the foundations of the earth were first exposed to my view, and where I first saw granite, limestone &c in those places, and those forms, where the ever working and all wonderful hand of nature had placed them. No Ben, it is impossible you can conceive my feelings and it is as impossible for me to describe them.

Naples: engraving after Turner.

Naples was the next stop, where a climb to the top of Mount Vesuvius during the night showed the volcano at its most dramatic, though a picnic provided a welcome respite (and a timeless image of British tourism):

We gained the summit by about half-past seven o'clock. During our climb upwards many beautiful views were highly enjoyed, and the evening light on the mountains and promontories was very fine. Some rain fell as we approached the top, which, being volatilised by the heat, made a much greater appearance of vapour than yesterday; added to which the fire was certainly stronger, and the smoke emitted in far greater quantity. The wind had changed since yesterday, but was still very favourable to our intention, and carried the smoke and vapour in a long black line over the hills to a great distance. It now became dark very quickly, and the flames appeared more and more awful (at one time enclosed in the smoke) and everything hid from our eyes; and then the flames flashing upwards and lighting through the cloud, till by a turn of the wind the orifice was cleared, and the dreadful place appeared uncovered and in all its horrors. The flames then issued forth in whirlwinds, and rose many yards above the mouth of the volcano. The flames were of a light red colour, and at one time, when I had the most favourable view of the mouth, appeared to issue from an orifice about three yards, or rather more, over.

Cloths were now laid on the smoking lava, and bread, chickens, turkey, cheese, wine, water, and eggs roasted on the mountain, brought forth, and a species of dinner taken at this place. Torches were now lighted, and the whole had a singular appearance; and the surrounding lazzaroni assisted not a little in adding to the picturesque effect of the scene. After having eaten and drunk, 'Old England' was toasted, and 'God save the King' and 'Rule, Britannia' sung; and then two very entertaining Russian songs by a gentleman (a native of that country) the music of which was peculiar and very touching.

Another natural phenomenon that fascinated young Michael was the light emitted by glow worms, and there are several mentions in his journal of rather gruesome experiments to find out how it worked. As always, the facts observed are meticulously documented. Indeed, several of Faraday's findings (such as that heat increases the brightness, but curtails the length of time the illumination lasts) give direct insight into what nowadays is called chemiluminescence:

This evening many glow worms appeared and, of four which I had put in a tumbler with green leaves, two shone very brightly. I separated the luminous part of one in full vigour from the body. It soon faded and in about ten minutes ceased to emit light, but on pressing it with a knife, so as to force the matter out of the skin, it again became luminous and continued to shine for two hours brightly. One I found on the floor, crushed unawares by the foot. I separated the luminous part of this insect and left it on paper. It shone with undiminished lustre the whole evening and appeared not at all to have suffered in its power of emitting light by the mixture and confusion of its parts, so that it appears to depend more upon the chemical nature of the substance than upon the vital powers of the animal, but, at the same time, it appears from the variations in splendour accompanied by motions in the living animal that it may be much influenced or modified by, or in some matter submitted to, the powers of the worm.

The light is very bright, sufficiently so to render the printing of a book very distinct, and I have often read my watch by it. It is of a delicate greenish tinge. The worm in its whole appearance resembles greatly a large woodlouse. The male is a fly, fully one-third smaller than the worm and emitting no light. The matter which appears to fill the hinder part of the body in the shining season is yellowish-white, soft and glutinous. It is insoluble in water or more in alcohol than in water. Heat forces out a bright glow, and then it becomes extinct; but if not carried too far, the addition of moisture after a time revives its power. The death of the worm seemed to have no immediate effect upon the illumination of the hinder part, and with respect to the length of time that it continued to shine afterwards, it seemed indifferent whether it was left on the body or taken off; but when extinct, exposure of the interior to air always caused a fresh emanation of light.

Supplementing the journal itself, several of Faraday's letters to relatives and friends in England give us his opinions about the life he was leading, and the national characters of the countries he passed through, the latter usually compared unfavourably with those of his follow countrymen. Faraday was nothing if not a patriot. In a letter to Robert Abbott, brother of his closest friend Ben, he sums up the many ways by which the English way of life and habit of thought surpasses, in particular, the French. It is very striking how little reference he makes, either in the journal or his letters, to the war against Napoleon, though doubtless the high opinion of England he professes to hear expressed on all sides emanated from those countries taking part in the Grand Alliance:

> I valued my country highly before I left it, but I have been taught by strangers how to value it properly, and its worth has been pointed out to me in a foreign land. In foreign nations alone it is that, by contrast, the virtues of England, her strength, and her wisdom can be perceived, and for those virtues Englishmen are considered every where as a band of brothers, actuated by one heart and one mind, and treading steadily and undeviating in the path of honour, courage and glory. All nations reverence them; every individual speaks highly of them; the English are respected, received, caressed everywhere for the character of their country: may she ever deserve that character: may her virtues still continue to shine in and illuminate the world: may her path never deviate from the point; but still proceeding in a straight line carry her still farther out in the view of the world, where her character meets with its just and honorable reward.

> On entering France, the dissimilarity between the inhabitants and the people of my own country was strong and impressive, and entered firmly in my mind. I have found the French people in general a communicative, brisk, intelligent, and attentive set of people; but their attentions were to

gain money, and for their intelligence they expect to be paid. Politeness is the general character of the people, a character which they well deserve; but the upper classes have carried it beyond the bounds of reason, and in politeness they lose truth and sincerity; their manners are very insinuating and kind, their address at once easy and free, and their conversation vivid and uninterrupted; but though it dwell for hours on the same subject, you can scarcely make out what the subject is: for it is certainly the most unfixed, most uninteresting, and unapplicable conversation I have met with.

Civilization seems to have taken different paths in the nations of Europe, towards the end of (or rather latter part of) her progress. At Paris, civilization has been employed mostly in the improvements and perfection of luxuries, and oftentimes in the pursuit has neglected the means of adding to domestic and private comfort; and has even at times run counter to it. In ornaments indeed, the Parisians excel, and also in their art of applying them; but in the elegance of appearance, utility is often lost, and English articles which have been formed under the direction of a less refined, but more useful judgement, are often eagerly preferred.

I, who am an Englishman and who have been bred up with English habits, of course prefer English civilization to the civilization of France, and think that my common sense has made the best choice; but every day's experience teaches me that others do not think so. Yet, though I have no right to suppose I excel all those who differ from me, I still am allowed the liberty of forming my own opinion.

Not mentioned explicitly in Faraday's journal, but an important ingredient in the experience he gained about social class and human relations during the continental tour, was the behaviour of Lady Davy. As we have seen already, the original plan called for Sir Humphry to be accompanied by his valet La Fontaine. Because of his last minute refusal to join the expedition,

Faraday was left to carry out some of the menial jobs that the valet would have occupied himself with, in addition to helping Davy in his capacity as scientific assistant. Davy himself came from a relatively humble background, and had no great need for a full time manservant; Lady Davy, on the other hand, had been used to more and her condescending manner became increasingly irksome to the young assistant, who (though of extremely humble origins himself) nevertheless felt that he had something more of the status of a scientific apprentice than a mere menial. In the first of several letters to Benjamin Abbott, in the autumn of 1814, frustration boils over, though at first expressed only in generalities:

> I have, Ben, sincerely observed that in the progress of things, circumstances have so worked together without my knowing how, or in what way, that an end has appeared which I could never have fancied, and which circumstances ultimately shew could never have been obtained by any plans of mine. I have found, also, that those circumstances which I have earnestly wished for (and which ultimately I have obtained) were productive of effects very different to those I had assigned to them, and were often times more satisfactory than ever a disappointment would have been. I have experienced, too, that pleasures are not the same when attained as when sought after, and from these things I have concluded that we generally err in our opinions of happiness and misery.

> In the first place then, my dear Ben, I fancy that when I set my foot in England I shall never take it out again, for I find the prospect so different from what it at first appeared to be that I am certain, if I could have foreseen the things that have passed, I should never have left London. In the second place, enticing as travelling is (and I appreciate fully its advantages and pleasures) I have several times been more than half decided to return hastily home, but second thoughts have still induced me to try what the future may produce, and now I am only retained by the wish of improvement.

I have learned just enough to perceive my ignorance and, ashamed of my defects in every thing, I wish to seize the opportunity of remedying them. The little knowledge I have gained in languages makes me wish to know more of them, and the little I have seen of men & manners is just enough to make me desirous of seeing more; added to which, the glorious opportunity I enjoy of improving in the knowledge of Chemistry and the Sciences continually determines me to finish this voyage with Sir Humphry Davy, but if I wish to enjoy those advantages, I have to sacrifice much and though those sacrifices are such as an humble man would not feel, yet I cannot quietly make them.

In Rome two months later, loneliness and alienation reach their peak and disgust with society boils over. Studying nature is seen as an escape from the enmeshments of base humanity:

Alone in a foreign country amongst strangers, without friends, without acquaintances, surrounded by those who have no congenial feelings with me, whose dispositions are opposite to mine and whose employments offend me, where can I look for pleasure but to the remembrances of my friends? At home I have left those who are dear to me from a long acquaintance, a congeniality of mind, a reciprocal feeling of friendship, affection and respect, as well as for their honour and their virtues. Here I find myself in the midst of a crowd of people who delight in deceiving, are ignorant, faithless, frivolous, and at second sight would be my friends. Their want of honour irritates me, their servility disgusts me and their impertinence offends me, and it is with a painful sensation I *think* of my friends when I remember I cannot do more.

Alas! how foolish perhaps was I to leave home; to leave those whom I loved and who loved me, for a time uncertain in its length, but certainly long and which perhaps may stretch out into eternity. And what are the boasted advantages to be gained? Knowledge, yes; knowledge, but what knowledge?

Knowledge of the world of men, of manners, of books and of languages: things in themselves valuable above all price, but which every day shews me prostituted to the basest purposes. Alas, how degrading it is to be learned when it places us on a level with rogues and scoundrels! How disgusting, when it serves but to shew us the artifices and deceit of all around! How can it be compared with the virtue and integrity of those who, taught by nature alone, pass through life contented, happy, their honour unsullied, their minds uncontaminated, their thoughts virtuous, ever striving to do good, shunning evil and doing to others as they would be done by. Were I by this long probation to acquire some of this vaunted knowledge, in what should I be wiser? Knowledge of the world opens the eyes to the deceit & corruption of mankind (of men), serves but to shew the human mind debased by the vilest passions of manners, points out the exterior corruption which naturally results from the interior of books. The most innocent occasions disgust when it is considered that even that has been debased by the corruption of many, and of languages serves but to shew in a still wider view what the knowledge of men and of manners teaches us. What a result is obtained from knowledge, and how much must the virtuous human mind be humiliated in considering its own powers when, at the same time, they give him such a despicable view of his fellow creatures.

Abbott replied at once, asking what the matter was, and elicited a more measured and explicit account of the reason for Faraday's anger, which, from the benefit of passing time, had cooled to a more judicious tone: Lady Davy was indeed at the heart of the problem. Still, it would appear that Michael was learning quickly how to combat her onslaughts in determined style:

Though it may appear somewhat consequential that I begin the letter with my own affairs, yet such is my intention at present. You found me in the last squabbling almost with all the world, and crying out against things which truly in themselves are excellent, and which, indeed, form the only distinction between men and beasts. I scarce know now what I said in that letter but I know I wrote it in a ruffled state of mind which (by the bye) resulted from a mere trifle.

I fancy I have cause to grumble, and yet I can scarcely tell why. If I approve of the system of etiquette and valuation formed by the world, I can make a thousand complaints, but perhaps if I acted influenced by the pure and unsullied dictates of common sense, I should have nothing to complain of and therefore all I can do is to give you the circumstances.

When Sir Humphry Davy first made proposals to me to accompany him in this voyage he told me that I should be occupied in assisting him in his experiments in taking care of the apparatus, and of his papers & books, and in writing and other things of this kind and I, conceiving that such employment (with the opportunities that travelling would present) would tend greatly to instruct me in what I desired to know, and in things useful in life, consented to go. Had this arrangement held, our party would have consisted of Sir Humphry and Lady Davy, the Lady's maid, La Fontaine (Sir H's valet) & myself. But a few days before we came off, La Fontaine, diverted from his intentions by the tears of his wife, refused to go and thus a new arrangement was necessary. When Sir H. informed me of this circumstance, he expressed his sorrow at it, and said he had not time to find another to suit him (for La Fontaine was from Flanders and spoke a little Italian as well as French) but that if I would put up with a few things on the road until he got to Paris (doing those things which could not be trusted to strangers or waiters, and which La Fontaine would have done) he would there get a servant, which should leave me at liberty to fill my proper station, and that alone. I felt unwilling to proceed on this plan but, considering the advantages I should lose and the short time I should be thus embarrassed, I agreed.

I should have but little to complain of were I travelling with Sir Humphry alone, or were Lady Davy like him, but her temper makes it often times go wrong with me, with herself & with Sir H. She is haughty and proud to an excessive degree and delights in making her inferiors feel her power. She wishes to roll in the full tide of pleasures such as she is capable of enjoying; but when she can with impunity (that is, when her equals do not notice it and Sir H. is ignorant of it) she will exert herself very considerably to deprive her family of enjoyments. When I first left England, unused as I was to high life and to politeness; unversed as I was in the art of expressing sentiments I did not feel, I was little suited to come within the observation (and under the power in some degree) of one whose whole life consists of forms, etiquette and manners. I believe at that time that she hated me, and her evil disposition made her endeavour to thwart me in all my views, and to debase me in my occupations. This at first was a source of great uneasiness to me, and often times made me very dull and discontented, and if I could have come home again at that time you would have seen me before I had left England six months. As I became more acquainted with the manners of the world, and those things necessary in my station, and understood better her true character, I learned to despise her taunts and resist her power, and this kind of determined conduct (added to a little polishing which the friction of the world had naturally produced in your friend) made her restrain her spleen from its full course.

Finally Sir H. has no valet except myself but, having been in an humbler station (and not being corrupted by high life) he has very little occasion for a servant of that kind, and 'tis the name more than the thing which hurts. I enjoy my original employment in its full extent and find few pleasures greater than doing so. At all event, when I return home I fancy I shall return to my old profession of Bookseller, for books still continue to please me more than anything else.

A fortnight later he wrote to another friend, Thomas Huxtable, in a rather world-weary tone; home was beckoning:

> ... As for me, like a poor unmanned unguided skiff, I pass over the world as the various and ever-changing winds may blow me, for a few weeks I am here, for a few months there, and sometimes I am I know not where; and at other times I know as little where I shall be. The change of place has, however, thrown me into many curious places and on many interesting things, and I have not failed to notice (as far as laid in my power) such things as struck me for their importance or singularity.

As the party began its journey back to England, Faraday reported a wry vignette of what might have been a nasty accident for an inexperienced horseman:

> Now though I am no rider, yet the circumstance must not be attributed to me alone that the horse and I were twice heels over head, but rather, it is a wonder that it did not happen oftener in nine miles. A tailor would have said that the horse was religious, and that it only did as other Italians do when they grow old and feeble, but that did not satisfy me and I would rather have had a beast that would have gone on orderly upon his legs.

With the trip nearing its end, two letters to his family give touching glimpses of Michael's feelings, first for his younger sister Margaret, and finally, for his mother. The didactic tone of an elder brother is evident in the way he passes on his own methods of self-improvement:

> I am pleased to hear that you go to school, and I hope that you have enough to do there. Your writing is not improved quite so much in one year as I expected it would be when I left home, but however it is pretty well. Your I's are most in fault. You must make them thus, *I I I I* with smaller heads. My questions about Rome and Naples I did not expect you could

answer, but I wished you to look into some book at school, or at Mr Riebau's or elsewhere, and give me the answers from them, at the same time fixing them in your memory. I gave them to you as lessons and I still hope you will learn them. I hope that you do not neglect your ciphering and figures, they are almost as necessary as writing and ought to be learned even in preference to French. Of this last you say nothing, but I suppose you still work at it. I will tell you my way of learning the words. When, in my grammar or in other books, I meet with a word (and that happens often enough) that I do not know, I first write it down on a fair sheet of paper, and then look in my dictionary for its meaning, and, having found it, I put it down also but on another part of the same sheet. This I do with every word I do not know very well, and my sheet of paper becomes a list of them, mixed and mingled together in the greatest confusion, English with French and one word with another. This is generally a morning's work. In the evening I take my list of words and my dictionary and, beginning at the top, I go regularly down to the bottom. On reading the words, I endeavour to learn their pronunciation, and if I cannot remember the meaning in the other language, I look in the dictionary and having found it, endeavour to fix it in my memory and then go to the next word. I thus go over the list repeatedly and on coming to a word which I have by previous readings learned, I draw a line over it, and thus my list grows little every evening and increases in the morning and I continually learn new and the most useful words from it. If you learn French and pursue this plan at home you will improve in it very quickly.

Joy at the approach to his native land shines from the final letter to his mother from outside England. It was written in Brussels on 16 April 1815. It may seem strange that neither in his journals nor in his letters does Faraday make more than passing reference to the momentous political events that were developing around the group of travellers as they criss-crossed the continent: the exile of Napoleon to Elba, his subsequent escape, and the

'Hundred Days' leading to the Battle of Waterloo. But for Michael these happenings were of little consequence beside the pleasure of rejoining his family at 18 Weymouth Street (near Baker Street in Central London) and Mr Riebau the bookseller, whose apprentice he had been so many years before. The travellers reached London on 23 April 1815: they had been away for nearly 19 months:

My very dear Mother,

It is with no small pleasure I write to you my last letter from a foreign country, and I hope it will be with as much pleasure you will hear I am within three days of England. Nay more, before you read this letter I hope to tread on British ground but I will not make too sure, lest I should be disappointed, and the sudden change and apparent termination of our travels is sufficient to remind me that it may change again. But, however, that is not at all probable and I trust will not happen.

I am not acquainted with the reason of our sudden return. It is, however, sufficient for me that it has taken place. We left Naples very hastily, perhaps because of the motion of the Neapolitan troops and perhaps for private reasons. We came rapidly to Rome; we as rapidly left it. We ran up Italy; we crossed the Tyrol; we stepped over Germany; we entered Holland, and we are now at Brussels and talk of leaving it tomorrow for Ostend; at Ostend we embark and at Deal we land on a spot of earth which I will never leave again. You may be sure we shall not creep from Deal to London and I am sure I shall not creep to 18 Weymouth Street and then—but it is of no use. I have a thousand times endeavoured to fancy a meeting with you and my relations and friends, and I am sure I have so often failed; the reality must be a pleasure not to be imagined, nor to be described. It is uncertain what day we shall get to London, and it is also uncertain where we shall put up at. You may be sure that my first moments will be in your company. If you have opportunities, tell some of my dearest friends, but do not tell everybody—that is, do not trouble yourself to do it. I am of no consequence except to a few, and there are but a few that are of consequence to me, and there are some whom I should like to be the first to tell myself—Mr Riebau for one. My thoughts wander from one to another, my pen runs on by fits and starts, and I should put all in confusion. I do not know what to say and yet cannot put an end to my letter. I would fain be talking to you but I must cease.

Adieu till I see you, dearest Mother, and believe me ever your affectionate and dutiful son.

M. FARADAY

Tis the shortest and (to me) the sweetest letter I ever wrote you.

Chapter 4

Way of Life and Work

Faraday's single-minded devotion to experimental science, surpassed only by his strong attachment to the Royal Institution, caused him to decline various prestigious posts that were offered to him. Having established his pattern of work and living at 21 Albemarle Street, he had no wish to resign it for any other career, even academic, and most certainly not administrative. University College, London was founded in 1827 and Faraday was approached by Dioysius Lardner, the Professor of Natural Philosophy and Astronomy about the possibility that he might become Professor of Chemistry. In the first of numerous letters of the same kind, turning down (as gently and politely as he could) the blandishments of colleagues, Faraday's gratitude for the life given to him by the Royal Institution is evident. He had also undertaken a project to improve the formulation of optical glass for telescopes under contract with the Royal Society and the Board of Longitude. This contract brought much needed finance to the Institution and, apart from anything else, Faraday did not wish his employer to be disadvantaged. The financial problems of the Royal Institution were continual and endemic from its foundation:

> You will remember, from the conversation which we have had together, that I think it is a matter of duty and gratitude on my part to do what I can for the good of the Royal Institution in the present attempt to establish it securely. The

Faraday's study (top) and the dining room (bottom) in the flat at the Royal Institution, from the watercolours by Harriet Moore.

Institution has been a source of knowledge and pleasure to me for the last fourteen years, and though it does not pay me in salary for what *I now* strive to do for it, yet I possess the kind feelings & good will of its Authorities & members, all the privileges it can grant (or I require) and moreover, I remember the protection it has afforded me during the past years of my scientific life.

These circumstances, with the thorough conviction that it is a useful & valuable establishment, and the strong hopes that exertions will be followed with success, have decided me in giving at least two more years to it, in the belief that after that time it will proceed well into whatever hands it may pass. It was in reference to this latter opinion, and fully conscious of the great opportunity afforded by the London University of establishing a valuable school of chemistry (and a good name that I have said to you & Mr. Millington) that, if things altogether had been two years advanced, or that the University had to be founded two years hence, I should probably have eagerly accepted the opportunity. As it is, however, I cannot look forward two years & settle what shall happen then.

Upon general principles only, I should decline making an engagement so long in advance, not knowing what might in the meantime occur: and as it is, the necessity of remaining free is still more strongly urged upon me. Two years may bring the Royal Institution into such a state as to make me still more anxious to give a third to it. It may just want the last & most vigorous exertions of all its friends to confirm its prosperity, & I should be sorry not to lend my assistance (with that of others) to the work. I have already (and to a great extent for the sake of the Institution) pledged myself to a very laborious & expensive series of experiments on glass, which will probably require that time (if not more) for their completion, and other views are faintly opening before us. Thus you will see that I cannot with propriety accede to your kind suggestion.

Personal gain was never high among Faraday's priorities; time to pursue his studies was. His time, as he says in the following letter to Percy Drummond, the Lieutenant Governor of the Royal Military Academy at Woolwich is his only estate; a thought he had already articulated much earlier, in one of his letters to Benjamin Abbott (p 12):

Allow me to thank you and the other gentlemen who may be concerned in this appointment for the good opinion which has induced you to propose it to me. I consider the offer as a high honour and beg you to feel assured of my sense of it. I should have been glad to have accepted or declined it, independent of pecuniary motives, but my time is my only estate, and that which would be occupied in the duty of the situation must be taken from what otherwise would be given to professional business.

The contract on optical glass had occupied the majority of Faraday's time for four years when, at the beginning of July 1831, he informed the Council of the Royal Society of his wish to be relieved of the work so as to devote more time to projects of his own devising. A substantial practical reason why he was able to relinquish the glass project was that he had accepted the task of giving courses of lectures on chemistry to cadets at the Royal Military Academy in Woolwich. The extra stipend was obtained with less expenditure of the commodity he most valued (time) than the laboratory work on glass. Within two months, the 'other subject' turned out to be his greatest discovery, electromagnetic induction:

With reference to the request which the Council of the Royal Society have done me the honor of making, namely, that I should continue the investigation, I should under circumstances of perfect freedom assent to it at once. But, obliged as I have been to devote the whole of my spare time to the experiments already described and consequently, to resign the pursuit of such philosophical enquiries as suggested themselves to my own mind, I would wish, under present circumstances, to lay the glass aside for a while, that I may enjoy the pleasure of working out my own thoughts on other subjects.

On 29 August 1831, the discovery that we now know as electromagnetic induction was made (see Chapter 8). Just three weeks later he wrote to Richard Phillips, the editor of the *Philosophical Magazine* to inform him about progress. It was indeed a 'good thing' that he had found: not a weed but a very big fish indeed:

> I am busy just now again on Electro-Magnetism and think I have got hold of a good thing but can't say; it may be a weed instead of a fish that after all my labour I may at last pull up. I think I know why metals are magnetic when in motion though not (generally) when at rest.

Work in the laboratory at the Royal Institution, and management of its day to day affairs, especially organising the Friday Evening Discourses, and preparing and delivering many series of Christmas Lectures, kept Faraday unremittingly busy. He had little taste for purely social entertaining, and rarely accepted invitations to soirées or dinners, as in the following case, from John Rennie, the civil engineer:

> I am greatly obliged to you for your invitation, but am unable to accept it. I am reluctantly forced to forgo my friends' company at table from January to June, and very rarely at other times dine out. In fact I am not a social man.

Exception could be made, however, for the Presidents of the Royal Institution and the Royal Society, as his titular superiors:

> I am very much obliged to you for your kind invitation, but am under the necessity of declining it because of a general rule which I may not depart from without offending many kind friends: I never dine out except with our Presidents, the

Deputation inviting Faraday to become President of the Royal Society, by E Armitage.

Duke of Sussex or the Duke of Somerset, whose invitations I consider as commands. Under these circumstances I hope you will accept my obligations to you, though I cannot accept your favour.

Faraday's unwillingness to be diverted from the tasks that devolved on him as Director of the Laboratories of the Royal Institution, and his desire to use whatever remaining time was left to him in carrying out original work in science, led him to decline election to high offices in the world of science such as the Council of the Royal Society, even when invited by the Secretary of the Society, the zoologist John George Children in 1832:

I have been honoured by the receipt of your letter requesting to know on the part of His Royal Highness the President and Council of the Royal Society whether I could accept the office of a Member of Council if recommended & elected for the coming years.

In reply I beg respectfully to decline the probability of election: not that I undervalue the character of that high and responsible office, or esteem lightly the favourable opinion which would recommend me to it. But the time I can spare from imperative duties is already so small that I am anxious to devote the whole of it to original investigation, and I entertain a hope that his Royal Highness the President and the Council will think that in doing so I am performing to the best of my abilities my duties to Science & the Royal Society.

Even the exercise of keeping abreast of developments in other areas of science was sacrificed in devotion to those fields (certainly wide enough!) where his own special interests lay. On receiving some articles on astronomy from George Biddell Airy, then (1832) at Cambridge, Faraday replied:

Many thanks for your kindness in sending me your excellent papers. I only wish I could understand them, so as to do your ability justice. But I have been convinced by long experience that, if I wish to be respectable as a scientific man, it must be by devoting myself to the unremitting pursuit of one or two branches only, making up by industry what is wanting in force.

Whilst a warm host to his colleagues on Friday evenings at the Royal Institution, Faraday was not willing to spend time on social calls—'stay at home and work hard' was the key to life in science. To the Secretary of the Linnean Society, Francis Boote, he wrote:

I had trusted we should have seen you during our Evenings and (failing that) I have been quite sorry I could not find time to call upon you; but I feel that the best answer I can make to your compliments is to stay at home and work hard.

A superficial notion that such a life was narrow and constrained would neglect the intellectual delights that came from understanding complex matters revealed by experiment as described in two letters, respectively to John Lubbock, the Treasurer of the Royal Society in 1833 and John Milkington, former Professor of Mechanics at the Royal Institution, who later became a Professor at William and Mary College, Virginia, in 1836:

> I have been so deeply engaged in experimental investigations of Electricity that I have not read a Journal (English or Foreign) for many months. My matter in fact overflows; the doors that open before me are immeasurable. I cannot tell to what great things they may lead and I have worked, neglecting everything else, for the purpose. I do not know whether Mathematical are like Experimental labours; if they are, you will have an idea of my toil but at the same time of my pleasure.

> Your accounts of your transits over the world, and changes in the position of your family, almost startle & shame me, for I feel as if I could have shewn none of the energy and perseverance which carries you through all these things. I have been here so long (three & twenty years) attached to the Royal Institution that I feel as if I were a limpet on a rock, and that any chance which might knock me from my position would leave me but little hopes of attaching myself anywhere again. So much for the habit, which is just as strong in matters of feeling as in matters of body.

Nevertheless, though professing shame to his friends that he was not ready to venture out into the world, there is no hint that the limpet would ever willingly detach itself from the rock of the Royal Institution, to which it had adhered so long.

After his experience of devoting nearly three years (from 1828–1831) to improving the formulation of optical glass, a project that he undertook at the request of Humphry Davy entirely for the finance that it brought into the Royal Institution, Faraday had no further desire to become involved in what nowadays would be called contract research. In evaluating Faraday's response to a request that he become involved in perfecting the manufacture of iron, it should be born in mind that he had already contributed to the advancement of metallurgy by his work on the hardening of steel. Though willing to lend his time and effort to matters that he considered of national importance (such as his work for Trinity House and on safety in mines) he was quite unwilling to take time from his own investigations to undertake industrial development. Thus he replied to an enquiry from John James Chapman about the possibility of translating some of his earlier work on alloys into an industrial process in quite uncompromising terms:

> When a set of Capitalists embark immense sums into such a trade as that of Iron for the purpose of manufacturing it, they soon (for their own sakes) get all the knowledge that is well ascertained, & add to it much more, experimentally, of their own. This early possession is a consequence of the close connection existing between such knowledge & the increase (or even safety) of their property.
>
> These causes have brought large manufacturers like that of iron into what might be called a discovery state: those concerned know all that has been ascertained, & apply it; & further improvements (of which there are no doubt plenty in the womb of time) can only be made by *further discoveries*.
>
> Now I really have no time or inclination to take up a manufacture like that of iron. A very important fact to a manufacture, & requiring much labour for its development, is very often of such a nature as to give no scientific reputation; and then, on the other hand, the remaining stimulus of interest is not with me; for if I improved, others would profit, & at the same time try to evade acknowledging the source of its improvement that they might not be expected to share their profits.
>
> I had enough of endeavouring to improve a manufacture when I gave all my spare time for nearly three years in working in glass. One such experiment in a man's life is enough. I think I might have made three or four philosophical discoveries in that time if I had pursued my

own thoughts & views instead of working for a committee on a trading matter.

As regards the organisation of the affairs of the Royal Institution, Faraday (the limpet on the rock) became increasingly conservative and dissociated from everyday matters, though himself responsible, as a younger man, for many innovations. However, for example in a letter in 1849 to Benjamin Brodie, surgeon at St George's Hospital, he was clear-sighted enough to see that it was he who stood in the way of change:

> Here things have reverted very much to their former state, I rather think perhaps fitly. The time was probably too soon for any change. But when such a one as myself gets out of the way, then new conditions, new men, new views, and new opportunities may allow of the development of other lines of active operation than those heretofore in service; and then perhaps will be the time for change.

Passing time also saw even greater withdrawal from sociable converse, and in 1851 he wrote to Edward Magreth, the Secretary of the Athenaeum:

> The progress of Old time, bringing with him in my case and in all others the usual effects, tends with me to the diminution of income, and therefore necessarily the diminution of pleasures depending upon it. One of the first of these which I am constrained to give up is my Membership at the Athenaeum. Will you therefore have the goodness to communicate to the Committee my resignation.

Through the 1850s Faraday's work pattern became even more episodic, with intense periods of activity (of increasingly brief duration) interspersed with intervals of rest and quiet relaxation in the country. He analysed his own difficulties in sustaining mental exertion with undiminished clarity in 1852 to Thomas Andrews, Professor of Chemistry at the Belfast Academical Institution, later to become the Queen's University:

I find in myself an illustration of one of the chapters of Dr. Holland's late volume on physiological subjects, namely, that on the *time* essentially required in mental operations. That time is now with me considerable in proportion to what it was naturally, & the consequence is that I can only hold my way in a quiet progression of things. When I am involved in rapid changes of thoughts or persons then I have to use extra exertion mentally, & then confusion & giddiness comes on. All this I forget when I have been in the country for a few weeks about doing nothing, & then I think myself as able as ever to race with others. But having come home and gone to work upon oxygen and a magnetic torsion balance for a little while, I find the old warnings coming on & I have to suspend my occupation.

Faraday had learned French during his continental tour with the Davys, but knew no other foreign language well; mathematical symbolism, too, was a language he never acquired, as in the following, addressed to P T Riess in 1855:

It was a very great pleasure to me to receive your kind letter; and written in such English as made me ashamed of my ignorance of the German language. I never cease to regret the latter circumstance; for I am aware of the great stores of knowledge in that language which would then be open to me in relation to my especial pursuits, and which some how or other the system of publication in our country almost entirely shuts out from me. I have several times within the last 15 years set about acquiring it, but a result over which I have no power, namely, a gradually failing memory, has on these occasions made the labour of head so great, that I have been obliged to refrain from such an endeavour, as also from many others. You gave me your book some time back. I looked at it eagerly; but both by its language & its mathematical developments (for the use of symbols requires memory) it was shut out from me; and so I placed it in our library, where I am very glad to find it is of great use to others.

Apart from the British Association, Faraday rarely took part in scientific meetings, though he recognised the need that they fulfilled in bringing colleagues together, not only to discuss and confront one another's ideas but to become acquainted on a more personal level. The advice to his junior colleague John Tyndall in 1855 on how to react to provocation remains sterling good sense for the present day scientific community:

> These great meetings, of which I think very well altogether, advance science chiefly by bringing scientific men together and making them to know and be friends with each other; and I am sorry where that is not the effect in every part of their course. I know nothing except from what you tell me, for I have not yet looked at the report of the proceedings; but let me, as an old man who ought by this time to have profited by experience, say that, when I was younger, I often misinterpreted the intentions of people, and found that they did not mean what at the time I supposed they meant, and further that, as a general rule, it was better to be a little dull of apprehension when phrases seemed to imply pique, and quick on the contrary when they seem to convey friendly feeling. The real truth never fails ultimately to appear, and the opposing parties are, if wrong, sooner convinced when replied to forbearingly than when overwhelmed. All I want to say is that it is better to be blind to the results of partisanship, and be quick to see good will. One has more happiness in oneself in endeavouring to follow the things which make for peace. You can hardly imagine how often I have been heated in private when opposed, as I have thought unjustly and superciliously, and yet have striven (and succeeded, I hope) in keeping down replies of the like kind, and I know I have never lost by it.

In the same letter the war then going on in the Crimea provides a metaphor for advance in science:

> The secret of magnetic action is like a Sebastepol at least in this point, that we have to attack it in every possible direction,

and make our approaches closer and closer on all the sides by which we can force access.

Faraday's reply to a request from a Scandinavian colleague, C Hansteen, to offer employment to one of his protégés, gives a succinct and vivid picture of the way the Royal Institution functioned in the 1850s, and of his own working methods:

> Our Institution is not like your Universities. It is a private establishment: the Government does nothing for us, and we have no opportunities of receiving students. We have but one assistant and he is an ordinary workman. I formerly gave lectures to which the public were admitted by payment, but do not deliver any now, except six, once a year at Christmas, to the juvenile connections of our members & subscribers.
>
> For the same reason I have never had any student or pupil under me to aid me with assistance; but have always prepared and made my experiments with my own hands, working & thinking at the same time. I do not think I could work in company, or think aloud, or explain my thoughts at the time. Sometimes I and my assistant have been in the Laboratory for hours & days together, he preparing some lecture apparatus or cleaning up, & scarcely a word has passed between us:—all this being the consequence of the *solitary & isolated* system of investigation, in contradistinction to that pursued by a Professor with his aids & pupils as in your Universities.

Finally, in the last decade of his life he summarises for his old friend Auguste de la Rive how he came to learn the rudiments of chemistry as a young man from a book by Mrs Marcet, and of electricity from the *Encyclopaedia Britannica*:

> Mrs. Marcet was a good friend to me, as she must have been to many of the human race. I entered the shop of a bookseller and bookbinder at the age of 13, in the year 1804, remained there 8 years, and during the chief part of the

Mrs Marcet, author of *The Elements of Chemistry*.

time bound books. Now it was in these books, in the hours after work, that I found the beginnings of my philosophy. There were two that especially helped me; the Encyclopaedia Britannica, from which I gained my first notions of Electricity and Mrs. Marcet's conversations on chemistry, which gave me my foundation in that science. I believe I had read about phlogiston etc in the Encyclopaedia, but her book came as the full light in my mind. Do not suppose that I was a very deep thinker or was marked as a precocious person. I was a very lively, imaginative person, and could believe in the Arabian nights as easily as the Encyclopaedia. But facts were important to me & saved me. I could trust a fact, but always cross examined an assertion. So when I questioned

Mrs. Marcet's book by such little experiments as I could find means to perform, & found it true to the facts as I could understand them, I felt that I had got hold of an anchor in chemical knowledge & clung *fast* to it. Hence my deep veneration for Mrs. Marcet, first as one who had conferred great personal good & pleasure on me, and then as one able to convey the truths and principles of those boundless fields of knowledge which concern natural things to the young, untaught, and enquiring mind.

You may imagine my delight when I came to know Mrs. Marcet personally; how often I cast my thoughts backward, delighting to connect the past and the present; how often, when sending a paper to her as a thank offering, I thought of my first instructress, and such like thoughts will remain with me.

I have some such thoughts even as regards *your own father:*, for when, later in life, I was first at the Royal Institution and then abroad with Sir H. Davy, your father was one of the very earliest, I think I may say *the first*, who personally, at Geneva, and afterwards by correspondence, encouraged and by that sustained me.

Chapter 5

Colleagues and Friends

Starting from the many encounters with European scientists that he made during his tour of the continent with Humphry Davy in 1813–1815, Faraday maintained consistent contact with a wide circle of acquaintances, both directly in the fields of science centred on his own fields of endeavour and with others who shared his love of learning. Among the latter was John Herschel, an astronomer whose work and reputation Faraday had long admired.

> I am glad you like my last experiments, and I have the more pleasure in receiving your commendation than that of another person, not merely because there are few whose approbation I should compare with yours, but for another circumstance. When your work on the study of Nat. Phil. came out, I read it (as all others did) with delight. I took it as a school book for philosophers, and I feel that it has made me a better reasoner, and even experimenter, and has altogether heightened my character and made me (if I may be permitted to say so) a better philosopher.

News about work in progress has always been part of the converse between scientists whose common interests ripen into personal friendship. An important part of such converse, however, has always been to preserve confidentiality. Passing on information about new results to others, before the work was

J F Daniell and Faraday having a discussion.

expounded to the wider scientific community through the medium of publications in learned journals, was (and is) a solecism, or worse, if the results are taken up and exploited without authorisation from their originator. James Forbes, Professor of Natural Philosophy at the University of Edinburgh received the following advice in 1832:

> You speak of the transmission of information and the lesson it affords to discoverers. But is it not very annoying that one may not talk of a matter to one's most intimate friend lest it

should be misinterpreted, or perhaps given to another or, as has happened to myself (in other cases, *not* the present), be actually stolen? But I must not allow my recollection to dwell on these things, although they sometimes almost induce me to give up the pursuit of science for, exalted & noble as it is in itself and its outward appearance, it frequently presents, to the private knowledge of him who pursues it, quite as much that is degraded & base.

Many scientists who become immersed in their chosen subject feel that their own attachment may ultimately outweigh the importance of the field they have devoted themselves to. From the perspective of more than 150 years on, few would categorise Michael Faraday as a bore, the word he uses to William Whewell, the Cambridge historian, in 1835:

> I had begun to imagine that I thought more about Electricity and Magnetism than it was worth, and so a notion was creeping over me that after all I was perhaps only a *bore* to my friends by the succession of papers I was bold to send forth, and not that successful labourer for science which I was striving to be. Perhaps you will think so, too, when I tell you I have the tenth series in print and waiting to reach you, but whatever you may think, I am resolved to take your last letter as encouragement to go on.

To seek unity among the diverse manifestations of natural phenomena was a consistent thread in Faraday's life. Work, however, veered from one field to another as he restlessly explored the links connecting the physical laws of electricity and magnetism with the diverse world of chemistry. To the German chemist Eilhard Mitscherlich in 1838 we find:

> I have been so electrically occupied of late that I feel as if hungry for a little chemistry; but then the conviction crosses my mind that all these things hang together under one law, and that the more haste we make onwards, each in his own path, the sooner shall we arrive, and meet each other, at that

state of knowledge of natural causes from which all varieties of effects may be understood & enjoyed.

While casting his own scientific net as wide as any experimental scientist before or since, Faraday was well aware that it was impossible (and probably undesirable) to try to maintain a detailed grasp of all the diverse topics brought to his attention by colleagues. In the matter of terrestrial magnetism, for example, which he subsequently studied closely, he wrote the following in 1839 to his friend Edward Sabine:

> You are indeed exceedingly kind in sending me your report, which I confess myself unworthy of, for I have not pursued the subject of terrestrial magnetism minutely enough to be conscious of its full value, and in reading it soon perceive how little I know of the matter. But to myself, I feel fully aware I should apply the principle of the division of labour and in the hope of doing any thing well, should do that only.

Reluctance to take part in social converse stemmed not only from unwillingness to be diverted from more serious or pressing matters, but from a fragile mental state that showed itself progressively in loss of memory. In fact, in the years on either side of 1840, from when the following extract from a letter to the eminent French chemist Jean-Baptiste André Dumas dates, he suffered what would nowadays be called colloquially a nervous breakdown, which kept him from both lecturing and sustained research:

> You know that I am a recluse & unsocial, and have no right to share in the mutual good feeling of Society at large, for the man that does not take his share of goodwill into the common stock has no claim on others. Such is not the case, I hope, from any cold or morose feeling in the heart, but from particular circumstances, amongst which are especially mental fatigue and loss of memory. Do not think, therefore, that I am unaffected by your kindness, of which I feel quite unworthy. It has disturbed my feelings the more as it was

quite unexpected, for knowing your high station in science, and seeing your value as a man, I did not think you would spare much thought for me after your return to Paris.

Faraday made up in his extensive correspondence with scientific colleagues on the continent for his unwillingness to travel and meet them in person. Switzerland, and Geneva in particular, had been a specially memorable part of his continental tour with the Davys, becoming clothed with almost magical status in his memory, as in the following in 1845 to Auguste de la Rive:

> Your kind invitation for the scientific meeting in August is very pleasant to the thought, but I dare not hope for such happiness. I long to see Geneva and Switzerland again but there are many things which come between me and my desires in that respect. I know the kindness of your heart and how far I may draw upon you if I come, and I thank you most truly, not only for the invitation you have sent me, but for all the favours you would willingly shew me. Do you remember one hot day (I cannot tell how many years ago) when I was hot and thirsty in Geneva, and you took me to your house in the town and gave me a glass of water and Raspberry vinegar? That glass of drink is refreshing to me still.
>
> Adieu, my dear friend. Remember me kindly to Madame de la Rive and (if I am not too far wrong in the collocation of thoughts and remembrance of past things) bring me to mind with one or two young friends who shewed me a doll's house once, and with whom I played on the green.

Despite the long years during which they had not seen one another, correspondence with Auguste de la Rive reveals a personal attachment of the deepest kind, suffused by religious spirit, as when de la Rive's wife died:

> I knew of your sad loss and had heard also of your personal illness and its very serious character: but I knew also that

Auguste de la Rive, Swiss chemist and friend of Faraday.

you had that within that might sustain you under such deep trials. Do not be discouraged. Remember, wait patiently. Surely the human being must suffer when the dearest ties are rent but, in the midst of the deepest affliction, there is yet present consolation for the humble minded which (through the power that is over us) may grow up and give peace and quietness and rest. Your letter draws me out to say so much, for I feel as if I could speak to you on account of something more than mere philosophy and reason. They give but a very uncertain consolation in such troubles as yours, and indeed nothing is more unsatisfactory to me than to see a mere rational philosopher's mind fighting against the afflictions that belong to our present state and nature— as, on the contrary, nothing is more striking than to see such afflictions met by the weakest with resignation and hope. Forgive me if my words seem to you weak and unfit for the occasion. I speak to you as I have felt, and as I still hope to feel to the end, and your affectionate letter has drawn me forth.

Discussion of mortality, and the different ways in which minds and bodies fail, gives way to pleasure at the marriage of de la Rive's daughter:

I often wonder to think of the different courses (naturally) of different individuals, and how they are brought on their way to the end of this life. Some with minds that grow brighter and brighter but their physical powers fail; as in our friend Arago, of whom I have heard very lately by a nephew who saw him on the same day *in bed and at the Academy*, such is his indomitable spirit. Others fail in mind first, while the body remains strong. Others fail in both together; and others fail partially in some faculty or portion of the mental powers, of the importance of which they were hardly conscious until it failed them. One may, in one's course through life, distinguish numerous cases of these and other natures; and it is very interesting to observe the influence of the respective circumstances upon the characters of the parties and in what way these circumstances bear upon their happiness. It may seem very trite to say that *content* appears to me to be the great compensation for these various cases of natural change, and yet it is forced upon me, as a piece of knowledge that I have ever to call afresh to mind, both by my own spontaneous & unconsidered desires and by what I see in others. No remaining gifts, though of the highest kind; no grateful remembrance of those which we have had, suffice to make us willingly content under the sense of the removal of the least of those which we have been conscious of. I wonder why I write all this to you. Believe me, it is only because some expressions of yours at different times make me esteem you as a thoughtful man and a true friend. I have often to call such things to remembrance in the course of my own self examination, and I think they make me happier. Do not for a moment suppose that I am unhappy. I am occasionally dull in spirits, but not unhappy. There is a hope which is abundantly sufficient remedy for that, and as that hope does not depend on ourselves, I am bold enough to rejoice in that I may have it.

I do not talk to you about philosophy, for I forget it all too fast to make it easy to talk about. When I have a thought worth sending you it is in the shape of a paper before it is worth speaking of; and after that it is astonishing how fast I forget it again; so that I have to read up again and again my own recent communications and may fear that as regards others. I do not do them justice. However I try to avoid such subjects as other philosophers are working at, and for that reason have nothing important in hand just now. I have been working hard but nothing of value has come of it.

Let me rejoice with you in the marriage of your daughter. I trust it *will be* as I have no doubt it *has been* a source of great happiness to you. Your son too, whenever I see him, makes me think of the joy he will be to you. May you long be blessed in your children and in all the things which make a man truly happy, even in this life.

Though brought together by a common interest in science (philosophy), 'when philosophy has faded away, the friend remains . . .':

I must write you a letter though I have nothing to say (i.e. nothing philosophical) but I hope to feel with you that when philosophy has faded away, the friend remains. Do not think that I cannot (and do not) rejoice in reading and understanding all that your vigorous mind produces, but for myself, I feel I have little or nothing to return, and though, when my sluggish mind is moved, I can think determinately and write decidedly, yet being once written I fall back into quietude, and leave what has been said almost uncared for or unthought of; and so it is that I do not teaze you in letters with much of my philosophic opinions.

Nearer to home, a cancelled visit to the Dean of Westminster, evokes the image of blowing soap bubbles together. Faraday was using them at the time in his lectures to young people, and later they provided an imaginative means to enclose gases whose magnetic properties he was studying:

> I shall be hard at work at the Trinity House tomorrow evening or I should have made an exertion to reach the Deanery. As it is, I shall not be able.
>
> My soap bubbles were all very good but my carbonic acid was too recently prepared, indeed only the moment before. I had learnt the lesson before but in the hurry of the moment forgot it again.
>
> I wish I could come tomorrow night that we might blow soap bubbles against each other. What a beautiful and wonderful thing a soap bubble is!

Correspondence was not only a question of sharing memories, again, to Dumas in 1849:

> I can hardly think your duty and high occupation, apart as it is from every link that can recall or relate to a remembrance of me, can ever leave you a moment for imagination to travel hitherward: but, whatever our different destinies in, and paths through, life; that you may be prosperous in your proceedings & happy in your heart and home is the earnest hope & wish of one who will never forget you.

But of the insisting of the primacy of proof by experiment (to the German physicist J Plücher):

> I want to establish in your mind very clearly that you must not think I deny all that I do not admit. On the contrary, I think there are many things which may be true, and which I shall receive as such hereafter, though I do not as yet receive them; but that is not because there is any proof to the contrary, but that the proof in the affirmative is not yet sufficient for me.

And to physicist Joseph Henry, at the far remove of the United States, comparisons of the effects of growing old:

> I wonder whether I shall ever see America—I think not. The progress of years tell, and their effect on me is to blot out many a fancy which in former days I thought might perhaps work up into realities—and so we fade away. Well, I have had (and have) a very happy life at home. Nothing should make me regret that I cannot leave it and indeed, when the time for decision comes, home always has the advantage. Mrs. Faraday wishes to be kindly remembered to you. We look at your face painted in light by Mayall and (dare I say it?) like He and nature together have made you look very comfortable, and I suspect that we have both altered much since we last saw each other. My wife mourns with half mimic, half serious countenance over my changes, and chiefly that a curly head of hair has become a mere unruly grisly mop. I think that is on the whole the worst part of the change that 60 years nearly have made.

While mutual pleasures in landscape are compared favourably to exhibitions (that of 1851 being in Faraday's mind):

> Your account of the country you have been through excites me far more than Palaces or Exhibitions. The beauties of nature are what I most enjoy. Scenery, and above all the effects of light and shadow—morning and evening and midday, or a storm or a cloudy sky. My predilection is for out of door beauties and just now, I and my wife have run away from London to the seaside to get quiet & rest. My head even now aches and I feel very weary.

John Tyndall, Professor of Natural Philosophy at the Royal Institution, frequently visited the Alps during the summer and maintained a chalet at Mürren. His observations on the behaviour of water and ice formed the substance of the Christmas Lectures on 'Glaciers' that he gave at the Royal Institution in 1871. What seems to have appealed to Faraday in Tyndall's letter from Switzerland

John Tyndall, Professor of Natural Philosophy at the Royal Institution.

was a combination of vividly evocative writing with explanation for the phenomena he was describing. So impressed was he that he sent it for publication in the *Times* newspaper:

> I might not have written to you again, but for the receipt of your letter by my wife, detailing the ascent of Monte Rosa, and the enormous indiscretion I have committed thereupon. What shall I say? I have sent it to the 'Times'. There, the whole is out. I do not know whether to wish it may appear tomorrow, or the next day, or not. If you should dislike it, I shall ever regret the liberty I have taken. But it was so interesting in every point of view, showing the life and spirit of a philosopher engaged in his cause; showing not merely the results of man's exertions but his motives and his nature—the philosophy of his calling and vocation as well as the philosophy of his subject, that I could not resist; and I was the more encouraged to do so because, from the whole character and appearance of the letter, it showed that it was

an unpremeditated reaction, and that you had nothing to do with its appearance, i.e. it will show that, if it should appear. How I hope you forgive me. Nobody will find fault with me, but you.

The German chemist Justus Liebig.

Faraday first met the German chemist Justus Liebig in 1837. They corresponded throughout their careers yet, towards the end of his life, Faraday's thoughts were of Liebig the man:

It seems very long since I wrote last, and I seem as if I had left a pleasure unenjoyed: but I have often thought of you, had thoughts even of seeing you; though ever as the proposed time drew near, things before unthought of grew into realities, and the dreams which seemed sometimes as lively as realities, passed away; and so it is with our life and so I suppose it ought to be. But the thoughts of you are pleasant; and my wife and I often think of the days at York, and then set too, to imagine what the years between then and now have done with you. I do not mean as to *progress, discovery,* and *fame* for that we know; but as to the personality of the *Man Liebig,* whose company and converse we enjoyed so much there, that it has left an enduring impression on my failing memory.

Another of Faraday's longest term correspondents was Christian Friedrich Schoenbein, Professor of Physics and Chemistry at the University of Basle from 1835–1852, and best known for discovering and investigating the chemistry of ozone. Indeed, the way in which this strongly oxidising gas is produced by electrolysing strong nitric acid solutions was the subject of their first letters. As time went by, their exchanges took on a more personal flavour, as, for example, in the year of revolutions in continental Europe, when Faraday expresses relief that events in Britain were not taking the same course:

> What a delight it is to think that you are quietly and philosophically at work in the pursuit of science, or else are enjoying yourself with Madame Schoenbein and the children amongst the pure and harmonious beauties of nature, rather than fighting amongst the crowd of black passions and motives that seem nowadays to urge men everywhere into action. What incredible scenes everywhere; what unworthy motives ruled for the moment, under high sounding phrases, and at the last, what disgusting revolutions! Happy are we here who have thus far been kept from these things, and hope to be so preserved in the future.

In 1853, when the spiritualist fad for table turning was at its height, Faraday took the opportunity of expressing to Schoenbein his strong abhorrence for such credulousness on the part of the public. He also made his views public in a letter to the *Times,* reproduced in Chapter 12:

> I have not been at work, except in turning the tables upon table turners. Nor should I have done that but that so

many enquiries poured in upon me that I thought it better to stop the inpouring flood by letting all know at once what my views and thoughts were. What a weak, credulous, incredulous, unbelieving, superstitious, bold, frightened, what a ridiculous world ours is, as far as concerns the mind of man. How full of inconsistencies, contradictions and absurdities it is. I declare that taking the average of many minds that have recently come before me (and apart from that spirit which God has placed in each) and accepting for a moment that average as a standard, I should far prefer the obedience, affections and instinct of a dog before it. Do not whisper this however to others.

Two years later, Schoenbein sent Faraday a book containing passages devoted to eulogising his life and work, which elicited the following self-deprecating comment, written from Hastings:

I cannot tell what sort of a portrait you have made of me. All I can say is that, whatever it may be, I doubt whether I should be able to remember it. Indeed I may say, I know I should not, for I have just been under the sculptor's hands, and I look at the clay and I look at the marble, and I look in the glass, and the more I look the less I know about the matter and the more uncertain I become. But it is of no great consequence; label the marble, and it will do just as well as if it were like. The imperishable marble of your book will surely flatter.

A further expression of friendship, when Faraday was 69, is eloquent of his personal feelings for Schoenbein:

After all, though your science is much to me, we are not friends for science sake only, but for something better in a man, something more important in his nature: affection, kindness, good feeling, moral worth; and so, in remembrance of these, I now write to place myself in your presence and, in thought, shake hands, tongues, and hearts together.

Chapter 6

Words for Things

Faraday's formal education ceased at the age of 13, in 1804, when he was apprenticed to the bookseller and bookbinder Mr Riebau. It is not surprising, therefore, that his knowledge of mathematics was rudimentary, and remained so throughout his life. Indeed, it is unlikely to have stretched much beyond simple arithmetic and though his notes contained geometrical diagrams (for example the lines of induction around magnetised bodies), there is no evidence that he was capable of drawing quantitative conclusions from them. Among the earliest of his protestations about his handicap in comprehending mathematical reasoning comes in a letter to the French physicist André-Marie Ampère in 1822. Still, he takes refuge in the primacy of experiment fact, regarding mathematics as a navigational compass of uncertain efficacy:

> I am unfortunate in a want of mathematical knowledge and the power of entering with facility into abstract reasoning. I am obliged to feel my way by facts closely placed together, so that it often happens I am left behind in the progress of a branch of science (not merely from the want of attention) but from the incapability I lay under of following it, notwithstanding all my exertions. It is just now so, I am ashamed to say, with your refined researches on electro-magnetism or electrodynamics. On reading your papers and letters, I have no difficulty in following the reasoning, but

still at last I seem to want something more on which to steady the conclusions. I fancy the habit I got into of attending too closely to experiment has somewhat fettered my powers of reasoning, and chains me down, and I cannot help now and then comparing myself to a timid ignorant navigator who (though he might boldly and safely steer across a bay or an ocean by the aid of a compass which in its actions and principles is infallible) is afraid to leave sight of the shore because he understands not the power of the instrument that is to guide him. With regard to electro-magnetism also, feeling my insufficiency to reason as you do, I am afraid to receive at once the conclusions you come to (though I am strongly tempted by their simplicity and beauty to adopt them).

Conversely, Faraday was intensely concerned with inventing and defining new words to encapsulate the new phenomena that his experiments were bringing to light. The ironic tone of the following description (in a letter of 1831 to the chemist Richard Phillips) of the coining of the word 'electrotonic', now quite vanished from the scientific vocabulary, should not be taken at face value:

> This condition of the matter I have dignified by the term *Electro-tonic. The Electro-tonic state.* What do you think of that? Am I not a bold man, ignorant as I am, to coin words? But I have consulted the scholars.

Despite his own propensity for coining words, it was not beyond Faraday to cast scorn on others who sought to do the same. The line between illumination and obfuscation through the use of technical language remains a narrow one for science to this day. William Whewell the Cambridge historian and classical scholar (to whom the following is addressed) frequently gave advice about etymology, clothing the new objects of science in classical garb:

> Your remarks upon chemical notation, with the variety of systems which have arisen with regard to notation,

William Whewell, Cambridge classicist and historian.

nomenclature, scales of proportional or atomic number etc, etc, had almost stirred me up to regret publicly that such hindrances to the progress of science should exist. I cannot help thinking it a most unfortunate thing that men who, as experimentalists and philosophers, are the most fitted to advance the general cause of science and knowledge should, by the promulgation of their own theoretical views under the form of nomenclature, notation or scale, actually retard its progress. It would not be of so much consequence if it was only theory and hypotheses which are thus treated, but they put facts (or the current coin of science) into the same limited circulation when they describe them in such a way that the initiated only can read them.

To another friend, William Nicholl, Faraday also sought advice about nomenclature, which he passes on in the following to

William Whewell, in 1834. The words 'electrolyte' and 'electrode' are still used to this day precisely in the sense in which Faraday defined them, but 'zetode' has not survived:

> I wanted some new names to express my facts in electrical science without involving more theory than I could help, and applied to a friend Dr Nicholl, who has given me some that I intend to adopt. For instance, a body decomposable by the passage of the electric current, I call an *'electrolyte'*, and instead of saying that water is *electro chemically decomposed* I say it is *'electrolyzed'*. The intensity above which a body is decomposed, beneath which it conducts without decomposition, I call the 'electrolytic intensity' etc, etc. What have been called the poles of the battery I call the *electrodes*. They are not merely surfaces of metal, but even of water and air, to which the term poles could hardly apply without receiving a new sense. *Electrolytes* must consist of two parts which, during the *electrolyzation*, are determined the one in one direction, the other in the other towards the electrodes or poles where they are evolved. These evolved substances I call *zetodes*, which are therefore the direct constituents of electrolytes.

Further suggestions from Whewell were gratefully accepted. Anode, cathode, cation and anion are still with us, and furnish the staple vocabulary of electrochemistry: the others (dexiode and skaiode) did not stand the test of time:

> All your names I and my friend approve of (or nearly all) as to sense and expression, but I am frightened by their length and sound when compounded. As you will see, I have taken *dexiode* and *skaiode* because they agree best with my natural standard East and West. I like Anode & Cathode better as to sound, but all to whom I have shewn them have supposed at first that by *Anode* I meant *No way*.

> I have taken your advice, and the names used are *anode*, *cathode, anions, cations* and *ions*; the last I shall have but little

occasion for. I had some hot objections made to them here, and found myself very much in the condition of the man with his son and ass who tried to please everybody; but when I held up the shield of your authority, it was wonderful to observe how the tone of objection melted away.

Two more words, used in the physical sciences to this day, that the world owes Faraday are diamagnetism and paramagnetism. The notion that lines of force (or, as we would say, magnetic induction) lie closer together inside a paramagnetic body than in free space, and further apart in a diamagnetic one, remains valid after 150 years. Faraday's defensiveness in not wishing his ideas to be bruited abroad until fully worked out is to be noted, as is the thought, expressed once again to William Whewell in 1850, that his ideas might shed some light on the earth's magnetism. He felt that paramagnetic oxygen in the atmosphere might be implicated; we know now that it is not:

> I have been driven to assume for a time, especially in relation to the gases, a sort of conducting power for magnetism. Mere space is Zero. One substance being made to occupy a given portion of space will cause more lines of force to pass through that space than before, and another substance will cause less to pass. The former I now call *paramagnetic* & the latter are the *diamagnetic*. The former need not of necessity assume a polarity of particles such as iron has when magnetic, and the latter do not assume any such polarity either direct or reverse. I do not say more to you just now because my own thoughts are only in the act of formation, but this I may say: that the atmosphere has an extraordinary magnetic constitution, and I hope and expect to find in it the cause of the annual & diurnal variations, but *keep this to yourself* until I have time to see what harvest will spring from my growing ideas.

Experiment, the diligent probing interrogation of nature, was ever the core of Faraday's approach to natural philosophy. Not only was this the case where particular theories fell to be tested, but additionally because the new facts uncovered might be valuable in themselves, even if the theory under test could not be conformed by them. A student of King's College, London, Frederick Oldfield Ward, received the following advice in 1834:

> I have no hesitation in advising you to experiment in support of your views, because, whether you confirm or confute them, good must come from your exertions.
>
> With regard to these views themselves, I can say nothing about them, except that they are useful in exciting the mind to inquiry. A very brief consideration of the progress of experimental philosophy will show you that it is a great disturber of pre-formed ideas.

Twenty years later A F Svanberg, a physics professor at Uppsala University was the recipient of quite similar advice:

> How wonderful it is to me, the simplicity of nature when we rightly interpret her laws, and how different the convictions which they produce on the mind, in comparison with the uncertain conclusions which hypothesis or even theory present.
>
> I am not sorry that you find some things unexpected or curious, or a little anomalous, for they serve to shew that there are more treasures to be obtained; and I see from your letter that you both know how to work them, and will work. The earnest ardent experimentalist is ever rewarded for his labour.

Faraday also tells his colleague John Tyndall in 1850 how working together on a given problem from different perspectives can also prove extremely fruitful:

> It is wonderful how much good results from different persons working at the same matter. Each one gives views and ideas new to the rest. When science is a republic, then it gains and, though I am no republican in other matters, I am in that.

Furthermore, also to Tyndall, one year later, an insistence on experiment as the ultimate arbiter between theories also has the effect of providing Faraday with an alibi for his inability to extract the deepest substance from mathematical reasoning:

> Mathematical formulae more than anything require quickness and surety . . . in receiving and retaining the true value of the symbols used, and when one has to look back at every moment to the beginning of a paper, to see what H or x or β mean, there is no making way. Still, though I cannot hold the whole train of reasoning in my mind at once, I am fully able to appreciate the value of the results you arrive at, and it appears to me they are exceedingly well established and of very great consequence. These elementary laws of action are of so much consequence in the development of the nature of a force which, like magnetism, is as yet new to us.

> I have far more confidence in the one man who works mentally and bodily at a matter than in the six who merely talk about it, and I therefore hope and (am fully persuaded) that you are working.

> Nature is our kindest friend and best critic in experimental science, if we only allow her intimations to fall unbiased on our minds. Nothing is so good as an experiment which, whilst it sets an error right, gives (as a reward for our humility in being reproved) an absolute advancement in knowledge.

Indeed, to James Clark Maxwell, the nineteenth-century's most eminent theoretical physicist, who took up and quantified Faraday's conception of fields of force around bodies, he makes a modest request in 1857 for the mathematically literate to translate their equations into concepts accessible to those not similarly equipped:

> There is one thing I would be glad to ask you. When a mathematician engaged in investigating physical actions and results has arrived at his own conclusions, may they not be expressed in common language as fully, clearly, and

definitely as in mathematical formulae? If so, would it not be a great boon to such as we to express them so—translating them out of their hieroglyphics that we might also work upon them by experiment? I think it must be so, because I have always found that you could convey to me a perfectly clear idea of your conclusions, which, though they may give me no full understanding of the steps of your process, gave me the results neither above nor below the truth, and so clear in character that I can think and work from them.

If this be possible, would it not be a good thing if mathematicians, writing on these subjects, were to give us their results in this popular useful working state as well as in that which is their own and proper to them?

Statements proved wrong by subsequent experiment are not a matter for shame, as when Faraday defends Tyndall in a letter to C Matteuchi (1855):

I think in that respect he is of my mind, that we are all liable to error, but that we love the truth, and speak only what at the time we think to be the truth; and ought not to take offence when proved to be in error, since the error is not intentional, but be a little humbled, and so turn the correction of the error to good account.

From fields of force around objects, as exemplified by magnetism, Faraday moved inexorably towards the view that a similar approach could be applied to other forces of quite different kind, especially gravity. The medium through which sound is transmitted is clear: the air. The notion that there is an 'ether' through which light is transmitted fell with the Michelson–Morley experiment some 30 years after Faraday wrote that science should also consider the mechanism by which the gravitational interaction is transmitted as opposed to the phenomenological laws propounded by Newton. In 1857 we find Faraday writing in these terms to Reverend E Jones, a minister at West Peckham in Kent:

The cases of action at a distance are becoming, in a physical point of view, daily more and more important. Sound, light, electricity, magnetism, gravitation, present them as a series. The nature of sound, and its dependence on a medium, we think we understand pretty well. The nature of light as dependent on a medium is now very largely accepted. The presence of a medium in the phenomena of electricity and magnetism becomes more and more probable daily. We employ ourselves, and I think rightly, in endeavouring to elucidate the physical exercise of these forces, or their sets of antecedents and consequents, and surely no one can find fault with the labours which eminent men have entered upon in respect of light, or into which they may enter as regards electricity and magnetism. Then what is there about gravitation that should exclude it from consideration also? Newton did not shut out the physical view, but had evidently thought deeply of it, and if he thought of it, why should not we, in these advanced days, do so too? Yet how can we do so if the present definition of the force, as I understand it, is allowed to remain undisturbed? Or how are its inconsistencies or deficiencies as a description of the force to be made manifest, except by such questions and observations as those made by me, and referred to in the last pages of your paper? I believe we ought to search out any deficiency or inconsistency in the sense conveyed by the received form of words, that we may increase our real knowledge, striking out or limiting what is vague. I believe that men of science will be glad to do so, and will even, as regards gravity, amend its description, if they see it is wrong. You have, I think, done so to a large extent in your manuscript, and I trust (and know) that others have done so also. That I may be largely wrong I am free to admit— who can be right altogether in physical science, which is essentially progressive and corrective? Still, if in our advance we find that a view hitherto accepted is not sufficient for the coming development, we ought, I think (even though we risk something on our own part), to run before and rise up difficulties, that we may learn how to solve them truly. To leave them untouched, hanging as dead weights upon our thoughts, or to respect or preserve their existence whilst they interfere with the truth of physical action, is to rest content with darkness and to worship an idol.

Chapter 7

Science at the Bench

Faraday's early career at the Royal Institution was very much taken up with chemical analyses brought from a wide variety of sources. He also began to give practical assistance to both private and government organisations. Principal among the latter in the 1820s was the project he undertook for the Admiralty to improve the quality of optical glass used in telescopes, but he had other illustrious clients, for example the pottery manufacturer Josiah Wedgwood, as witness his laconic note of February 1819:

Clays
No. 1. Cornwall Clay—dried

Silex	53.6
Alumine	45.6
Iron oxide	.4
	99.6

No. 3. Flintshire Clay—dried

Silex	59.3
Alumine	40.
Iron oxide	.3
	99.6

Mr Faraday has to apologise for the delay of these analyses but workmen in the Laboratory have retarded the usual operations there.

Faraday's laboratory: watercolour by Harriet Moore.

A month later, in sending another batch of analyses to Wedgwood, Faraday took advantage of the opportunity to enquire about the design of furnaces used to make pottery. He needed the information to help him with a project he was involved in to investigate the hardening of steel:

> Mr Faraday sends Mr Wedgwood the analyses of the clays and would be obliged if Mr Wedgwood could at a convenient opportunity give him short notices of their localities, times of discovery and relative utility.

> Mr Faraday takes the liberty of asking also whether Mr Wedgwood has any means which are not secret of ascertaining the heat of excellent furnaces. Mr Faraday is engaged in some experiments carried on at very high temperatures and feels curious on this point.

In 1820, Mr Stodart, who kept a shop selling cutlery and razors in the Strand commissioned the Royal Institution to analyse, and attempt to reproduce, a specially hard steel imported from India called Wootz. Faraday was asked to work on the problem, and after designing and building several furnaces he discovered that the secret was to add small amounts of heavy metals such as rhodium or silver to the melt. The results of two years experiments are summarised in a letter to his old friend Charles-Gaspard de la Rive, who he had first met in Geneva during his continental tour with Davy in 1813–1815. Faraday's experimental ingenuity is well to the fore, for instance in making crucibles that withstood the high temperatures needed, as also are his powers of observation. The pattern formed on the metal surface after etching by dilute acid (now known to be due to grain boundaries) reproduced that of the genuine imported product, the 'damask' sheen being the same as that of Damascus steel. Notice, too, the observation that silver is volatile, as well as the glancing mention that titanium could not be reduced to the metallic state. In fact titanium metal was not made until 1932, by reducing the oxide with aluminium:

It is possible you may have observed an analysis of Wootz or the Indian steel published in one of our Journals some time since. I could, at that time, find nothing in the steel besides the iron & carbon but a small portion of the earths or (as I presume) their metallic bases. On the strength of this analysis we endeavoured to demonstrate the particular nature of Wootz synthetically by combining steel with these metallic bases and we succeeded in getting alloys which, when worked, were declared by Mr. Stodart to be equal in all qualities to the best Bombay Wootz. This corroboration of the nature of Wootz received still stronger confirmation from the property possessed by the Alloys in common with Wootz, namely their power of yielding damasked surfaces by the action of acids. When Wootz is fused & forged, it still retains so much of the crystalline structure as to exhibit (when acted on by very weak sulphuric acid for some time) a beautiful damasked surface. This we have never yet seen produced by pure steel but it *is* produced in our imitation of wootz or alloys of steel with the metal of alumine.

Perhaps the very best alloy we have yet made is that with rhodium. Dr. Wollaston furnished us with the metal, so that you will have no doubts of its purity and identity. One and a half per cent of it was added to steel, and the button worked. It was very malleable but much harder than common steel & made excellent instruments. In tempering the instruments, they required to be heated full 70 degrees F. higher than is necessary for the best cast steel, and from this we hope it will possess greater degrees of hardness and toughness. Razors made from the alloy cut admirably.

You cannot imagine how much we have been plagued to get crucibles that will bear the heat we require and use in our experiments. Hessian, Cornish, pipe-clay crucibles all fuse in a few minutes if put into the furnace singly, and our only resource is to lute two or three, one within another together, so that the whole may not fuse before our alloy has had time to form in the centre. I have seen Hessian crucibles become so soft that the weight of 500 grains of metal has made them swell out like a purse, and the upper part has fallen together in folds like a piece of soft linen, and where three have been put together, the two outer ones have, in less than half an hour, melted off & flown down into the grate below.

From these circumstances you will judge of the heat we get, and now I will mention to you an effect which we obtain (and one which we can't obtain) both of which a little surprised us. The positive effect is the volatilization of silver. We often have it in our experiments sublimed into the upper part of the crucible and forming a fine dew on the sides and cover, so that I have no doubts at present on the volatility of silver, though I had before. The non effect is the non reduction of titanium. We have tortured Menachanite, pure oxide of titanium, the carbonate etc, in many ways in our furnace but have never yet been able to reduce it (not even in combination with iron) and I must confess that now I am very sceptical whether it has ever been reduced at all in the pure state.

Now I think I have noticed the most interesting points at which we have arrived. Pray pity us that, after 2 years experiments, we have got no farther, but I am sure if you knew the labour of the experiments you would applaud us for our perseverance at least.

Chapter 8

Leaves from a Laboratory Notebook

All practising scientists carrying out experiments keep laboratory records, usually in the form of a day to day account of experiments performed, with technical details such as the weights of materials used, and the experimental conditions. Faraday was no exception to this rule, but his notebooks are probably more extensive than those of any other great scientist. They are also notable for existing in an unbroken series from his first arrival as Humphry Davy's assistant right up to the end of his life in the laboratory. What makes them truly unique, however, is the fact that they remain to this day within the same building at 21 Albemarle Street where they were written.

Faraday's handwritten notebooks, which in printed transcript run to seven volumes, have long been of interest to historians and philosophers of science because of the extraordinarily direct insight they give into the way his thinking developed (as one would nowadays say) 'in real time'. They are also remarkable in the amount of detail that they give about the design and setting up of experiments, interspersed with comments about their outcome and thoughts of a more philosophical kind. All are couched in plain language, with many vivid phrases of delightful spontaneity. However, because Faraday was here communing with himself, many take the form of notes, and introduce words and phrases not familiar to a general audience (or archaisms that have fallen out of use). Consequently, only a brief and

Faraday's laboratory seen through the doorway: from a watercolour by Harriet Moore.

rather exiguous selection of extracts is given here, to provide some flavour of the whole. Nevertheless they encompass the first account of some of his most remarkable discoveries, such as electromagnetic induction, and contain many memorable phrases and insights.

The earliest notebooks contain rapidly written observations, not only of manipulations carried out in the laboratory, but descriptions of natural phenomena that took his eye and struck his imagination. A view of the dome of St Paul's shadowed on the clouds, for example, is described in beautifully clear prose. The companion who saw it with him was almost certainly his wife Sarah:

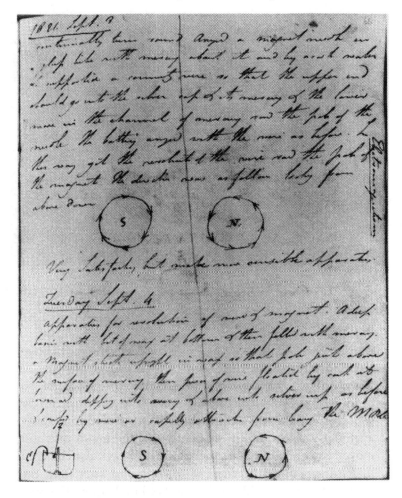

A page from Faraday's experimental diary.

1827. OCTR. 6TH.

A beautiful areal phenomenon observed about St. Paul's Church, from the shadow of the dome and the part above, cast on very thin clouds moving at that height. The moon at the full and rising. On looking at St. Paul's cross from Ludgate Hill, about Stationers' Court and the opposite side,

Faraday's laboratory, showing him at work: watercolour by Harriet Moore.

a stream of darkness seemed to issue from the part above the dome and, expanding, seemed to pass over the head of the spectator. By moving a little to the north or south, so as to get under the edge of the shadow it was exceedingly well defined and distinct. Each of the turrets in front threw a similar shadow, but more faintly. As the moon rose and moved towards the south, the shadows or rays also changed their directions, and at last they were best seen from the corner of Ludgate Hill and St. Paul's Church Yard. The mist or cloud was very faint, for the stars could be well seen. Whilst looking at it my companion thought she saw a black ray in another direction. This, however, proved to be the clear space between one thin cloud and another, and watching this and by tracing the motion of the cloud in the wind, we were able to account for the increase and diminution in strength of the shadow of the church spire as the cloud came up and afterwards passed over. The effect was very beautiful. Many persons went away fully convinced that rays of darkness were issuing from the Church. Time about 8 o'clock.

Laboratory notebooks (and Faraday's were no exception) often contained sketch drawings of apparatus. In the experiment he carried out on 30 May 1831 his aim was to see whether an electric current could be induced in a wire by subjecting it to a temperature gradient. The experiment was most ingeniously set up, as the diagram shows, but the result was negative:

1831. MAY 30TH.

A silver wire ring (a) was so put into a vessel of water (b) that a part could project over the edge and be heated by a spirit flame (c) whilst a little magnetic needle (d) suspended by a long delicate silk fibre hung over the top. The ring was then moved in the direction of the arrows, so that each part became hot and was then cooled suddenly in the water. No effect on the needle was observed, nor when the needle was perpendicular to the plane in which the wire moved. The expt. was made to ascertain whether, as the heat travelled from particle to particle, any electricity was put in motion.

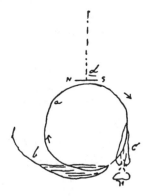

Faraday's sketch of apparatus for determining whether an electric current was induced by a temperature gradient.

29 August 1831 was one of the most momentous days in Faraday's career: the disturbance of the magnetic needle placed near coil B that he observed at the instant when the battery was connected or disconnected to coil A was evidence that an electric current had been *induced* in B by passing a current through A. The effect is called *electromagnetic induction*. Only the iron ring connected A with B, and the current was only induced in coil B when the current in A was changing. The phenomenon that Faraday describes is fundamental to the electrical generating and distribution industry. It is the basis of the transformer and the dynamo. It seems that Faraday was aware that his work this day was setting him off on a new and fertile pathway in science, because he began a new system for numbering the paragraphs in his notebook, starting again from number 1. Ever afterwards, up to his last recorded experiment 31 years later, he maintained such a numbering scheme:

AUG. 29TH, 1831

1. Expts. on the production of Electricity from Magnetism, etc. etc.

2. Have had an iron ring made (soft iron), iron round and 7/8 inches thick and ring 6 inches in external diameter. Wound many coils of copper wire round one half, the coils being separated by twine and calico—there were 3 lengths of wire each about 24 feet long and they could be connected as one

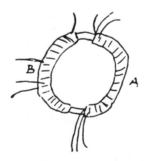

Faraday's first induction coil.

length or used as separate lengths. By trial with a trough each was insulated from the other. Will call this side of the ring A. On the other side but separated by an interval was wound wire in two pieces together amounting to about 60 feet in length, the direction being as with the former coils; this side call B.

3. Made the coil on B side one coil and connected its extremities by a copper wire passing to a distance and just over a magnetic needle (3 feet from iron ring). Then connected the ends of one of the pieces on A side with battery; immediately a sensible effect on needle. It oscillated and settled at last in original position. On *breaking* connection of A side with Battery again a disturbance of the needle.

Apparatus for determining the electrical conductivity of silver sulphide.

Amongst the myriad phenomena that Faraday can be credited with discovering, that of semiconductivity is not mentioned very often, probably because he did not publish it. It was known by 1833, when he carried out the experiment on 'sulphuret of silver' (what would nowadays be called silver sulphide), that the electrical resistance of metals increases when they are heated. Fusing some silver sulphide into a bent tube with two pieces of wire to connect it to a battery, Faraday found that the current he could pass through it actually increased as it became hotter, that is, its resistance went down. The effect (as we now know, and as Faraday carefully showed) does not come from any chemical reaction inside the material: it is purely electronic, and indeed forms the basis of the modern day electronics industry:

21ST FEBY. 1833.

316. *Sulphuret of Silver—very extraordinary.* At first on piece

of glass flask in air, but afterwards in tube, fused into its place in fire.

317. When all was cold conducted a little (by galvanometer) and if quite cold at first conducting power did not increase. But if battery current strong or if sulphuret continued to increase in conducting power, ... *The heat rose as the conducting power increased* (a curious fact), no other source of heat than the current being present. Yet I do not think it became high enough to *fuse the sulphuret*. The whole passed whilst in the solid state. The hot sulphuret seems to conduct as a metal would, and could get sparks with wires at the end and a fine spark with charcoal.

318. The sulphuret when hot seems to me very much to resemble the metals in their usual state as to electrical relations, a conductor, solid, not decomposable, etc. etc. etc.

319. If the sulphuret cooled then returned to first state and then all these effects could be obtained again. Hence no permanent change inside.

After discovering electromagnetic induction, Faraday devised many other arrangements of coils carrying current, but then took the further imaginative step of moving a bar magnet through a coil of wire, whereat a current also flowed. The 25th August 1834 was truly the genesis of the electric power industry: the simple apparatus he describes in clear detail here (and which can be seen in the Museum at the Royal Institution) was the first to convert mechanical energy directly into electrical energy—the dynamo:

25 AUG. 1834.

1933. To-day procured the Electric spark from a Magnet directly, ... The arrangement is here shewn in section.

A, a cylinder magnet 10 inches long, 8/10 in diameter; it would not lift more than 2oz. at either pole. B a pasteboard cylinder in which A could move freely. C,C, two collars of pasteboard backed by two bungs D,D, which were fastened by sealing wax to the cylinder B. E, about 6 feet of copper

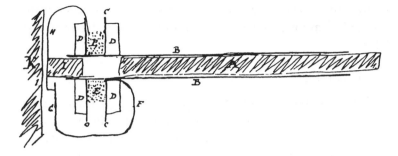

The first dynamo.

wire covered with white silk and coiled between the collars, C,C. F, one end of the coil soldered to a copper plate G. H, the other end of the coil, bent so as to rest upon the plate G and bear against it with a little spring. The wire and plate where they touched were amalgamated. I, a piece of wood which, when urged forward by the magnet A, would separate the wire H from G at the most favourable moment of time. K, a wall to stop the magnet.

1934. When A, being partly out as in the figure, was driven quickly up the cylinder, it opened H from G before its motion was concluded, and the spark was seen there. The Electricity here is much more directly from the magnet than in the usual way of procuring a spark.

Faraday gave his name to the electrochemical constant, the unit of electrical capacitance, magneto-optical rotation, and also to the 'Faraday cage', the first example of which (12 feet across!) he built in the Royal Institution lecture theatre. The whole surface of the cage (in this case a cube) was criss-crossed with wires, and the idea was to see, first of all, whether the faces, edges and corners were charged to the same extent when connected to a generator and, second, what the electrical charge might be inside the enclosure. For the second experiment Faraday himself got inside the cage, armed with an electrometer. The result (namely, that inside an enclosure formed by conducting material the potential is constant) is widely exploited to this day to protect sensitive electronic components from randomly fluctuating electromagnetic fields:

15 JANY. 1836

2808. Have been for some days past engaged in building up a cube of 12 feet in the side. It consists of a slight wooden frame, constituting the twelve linear edges, held steady by diagonal ties of cord; the whole being mounted on four glass feet, $5\frac{1}{2}$ inches long, to insulate it. The sides, top and bottom are covered in with paper. The top and bottom have each a cross framing or tying of copper wire, thus: which, with the diagonals of cord, support the two large sheets of paper which cover them in, the copper wire also serving to feed the paper surface with electricity. The framings at the top and bottom, of copper wire, are connected by copper wires passing down the four corner uprights; and a band of wire also runs round the lower edge of the cube. The sheets of paper which constitute the four sides have each two slips of tin foil pasted on their inner surface, running up 3/4 of the height; and these are connected below with the copper wire so that all the metallic parts are in communication. The edges of the side sheets are fastened here and there by tacks or paste to the wooden frame at the angles, so as to prevent them flying out and so giving irregular dispersion of the electricity. The whole stands in the Lecture room, one of the lower edges being within 5 inches of the third seat (on which the feet rest), and the opposite lower edge being sustained on stools and

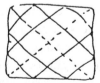 

Two sketches of the first Faraday cage.

blocks, about 4 feet from the ground. The chandelier hangs nearly opposite the middle of the face of the cube at this side, being about $2\frac{1}{2}$ feet from it.

2809. The cube rises in the middle of the room above the level of the bottom of the gallery, and appears of enormous size; it holds of course 1728 cubic feet of air.

2810. Connecting this cube by a wire with the Electrical machine, I can quickly and well electrify the whole.

2812. Proceeded to experiment with this cube, as follows. The large brass carrying ball was used, being suspended by a white silk thread to the end of a long glass rod (flint), in order to avoid (as much as might be) the inductive effect due to opposing the hand and body to the part from where a sample was to be taken. The machine and cube were retained, connected by a wire. Three turns of the machine were given whilst the carrier ball was held in contact with the part to be tested. The ball was removed from contact before the machine stopped—the conductor and cube discharged and the ball carried to the delicate electrometer (retained in a sheltered safe place so as not to be affected either by the machine or cube), and examined as to its charge and the degree acquired.

2813. A small charge was found on the middle of a face, as clear as possible under the circumstances from inductive influence. A stronger charge, by much, was obtained at the middle of an edge; and the strongest at the solid angle or corner.

2852. I now went inside the cube, standing on the stool and Anderson worked the machine until the cube was fully charged and he continued working the machine. I could by no appearance find any traces of electricity in myself or the surrounding objects. I could not affect the gold leaf electrometer within. But if I brushed it with flannel it was excited in the usual way.

2858. In fact the electrification without produced no

Faraday's magnetic laboratory: watercolour by Harriet Moore. His 'great electromagnet' is under the table.

consequent effects within, other than what belong to any open chamber.

From the intimate connection between magnetic fields and electric current flow revealed by the experiment of 29 August 1831 it became a central tenet of Faraday's world picture, and a guiding principle of much of his laboratory work after 1840, that an underlying hidden connection between all the forces of nature lay waiting to be uncovered by the patient and resourceful experimenter. Further dramatic confirmation of his belief came on 13 September 1845 when, in an experiment that ranks in significance at least as high as that of August 1831, he found that the polarisation of a light beam was affected by a magnetic field. The light beam was passed through a piece of heavy glass between the pole pieces of an electromagnet: 'heavy' glass because he had some pieces left from his work on optical glass in the 1820s. It was a good thing that he did, because had he used ordinary window glass, for example, the effect would probably not have been seen, since it is many times smaller. Though he could not have known it at the time, Faraday's result was not entirely serendipitous, however, because the electronic characteristics that give heavy (i.e. lead) glass a high refractive index, and hence make it suitable for telescope lenses, are the same ones that give it what nowadays we would call a high specific magnetic rotation:

7504. A piece of heavy glass (7495) which was 2 inches by 1.8 inches, and 0.5 of an inch thick, being a silico borate of lead, and polished on the two shortest edges, was experimented with. It gave no effects when the same *magnetic poles* or the *contrary* poles were on opposite sides (as respects the course of the polarized ray)—nor when the same poles were on the same side, either with the constant or intermitting current—BUT, when contrary magnetic poles were on the same side, there was an *effect produced on the polarized ray*, and thus magnetic force and light were proved to have relation to each other. This fact will most likely prove exceedingly fertile and of great value in the investigation of both conditions of natural force.

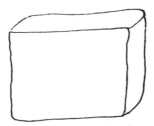

Heavy glass block.

7605. I used our strongest horseshoe magnet, and setting up a piece of heavy glass, approached the poles of the magnet and withdrew them again; then reversed the direction of the poles, approached them and withdrew them; and so on for some time.

7606.   There was a distinct influence upon the colour of
the feeble image in these two positions, proving the effect
of the magnetic curves of ordinary magnets, and also the
production of rotation in two directions. The influence was
very feeble, but still distinct enough for me to find out, by
its apparent contradiction in direction with the effects of the
electro magnets, that I had mistaken in them one pole for
the other in certain of the experiments, and it led to their
repetition and rectification.

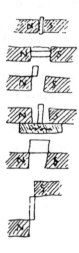

Arrangement of magnet pole pieces with respect to the glass block.

After a few more days experimenting, Faraday was fully
convinced, and brought his account to a triumphantly rhetorical
conclusion:

30 SEPTR. 1845.

Considering the nature of the relation between Mag. and
Electric forces, I think there must be some effect produced
which stronger magnets and other forms of apparatus and
the progress of our knowledge will enable us hereafter to
develop.

7718. Still, I have at last succeeded in *illuminating a magnetic curve* or *line of force* and in *magnetising* a ray of light.

15 NOVR. 1845.

8251.   Now we find *all* matter subject to the dominion of Magnetic forces, as they before were known to be to Gravitation, Electricity, cohesion.

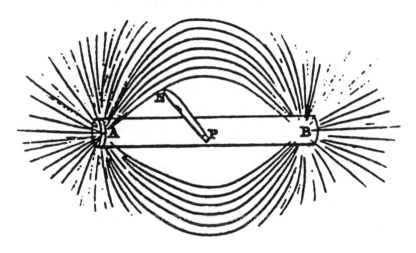

Faraday's first published drawing of the lines of magnetic flux around a bar magnet.

As the previous extract from his diary portends, the final frontier in Faraday's quest for connection between the various forces of nature was gravitation. It occupied much of his energy and experimental ingenuity from the end of the 1840s till his retirement. That such a search still goes on in the 1990s is not to say (yet) that it is in vain, but all attempts to detect the influence of gravitational fields on magnetism proved negative. Rehearsing a list of possible reasons for the negative outcome of one set of experiments, led Faraday to one of his most vivid (and frequently quoted) phrases.   But many who believe that 'nothing is too wonderful to be true' ignore the pragmatic tone of the rest of the sentence: experiment was forever to be paramount:

19 MAR. 1849

10040. ALL THIS IS A DREAM. Still, examine it by a few experiments. Nothing is too wonderful to be true, if it be consistent with the laws of nature, and in such things as these, experiment is the best test of such consistency.

In approaching gravitation as the last great subject of his work, Faraday was fully conscious of the awe-inspiring nature of what he was taking on:

25 AUG. 1849.

10061. I have been arranging certain experiments in reference to the notion that Gravity itself may be practically and directly related by experiment to the other powers of matter, and this morning proceeded to make them. It was almost with a feeling of awe that I went to work, for if the hope should prove well founded, how great and mighty and sublime in its hitherto unchangeable character is the force I am trying to deal with, and how large may be the new domain of knowledge that may be opened up to the mind of man.

Sadly, the 'extraordinary results' he was seeking were not to emerge:

25 AUG. 1849.

10112.   If there should be any truth in these vague expectations of the relation of Gravitating force, then it seems hardly possible but that there must be some extraordinary results to come out in relation to celestial mechanics—as between the earth and the moon, or the Sun and the planets, or in the great space between all gravitating bodies. Then indeed, Milton's expression of the Sun's magnetic ray would have a real meaning in addition to its poetical one.

Meanwhile, in parallel with the attempt to bring gravity into correlation with other forces, work on the magnetic properties of many substances continued apace. For instance, to observe whether gases like oxygen and nitrogen were paramagnetic or diamagnetic (that is, whether they were attracted towards, or repelled from, regions of high magnetic field) a most ingenious method was devised of encapsulating the gases inside soap bubbles and then holding them between the pole pieces of a magnet:

24 JUNE 1850.

10860.   I have endeavoured to ascertain the relation of different gases to each other in the magnetic field by blowing a bubble of gas at the end of a delicate pipe, and then submitting it to the magnetic force in another gas. A glass tube was drawn out to about the size and form of the drawing and fitted at the thicker end to a cap, stop cock and bladder to supply by pressure any given gas.   A little thin soap suds was made at the moment by putting a few cuttings of soap into a little distilled water and shaking them for an instant. Old soap suds or that made with warm water is not so good as this, being thicker and stringy. Then after drawing the soap suds up into the tube and also washing the outside of the termination with it, it was easy to make soap bubbles, beautiful in form and more or less ballasted by water beneath, and very mobile in a pendulous fashion because of the smallness of the neck where they adhered.

Likewise, arrays of iron filings, aligned in the lines of force generated by pairs of bar magnets held together in different

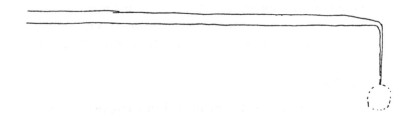

Tube for blowing bubbles containing gases.

arrangements, yielded beautiful pictures of the variation in field strength.

Not all experiments, however carefully and ingeniously designed, were so successful. In the following extract, from 4 August 1852, we catch a glimpse of the frustration that so often attends experimental science. Characteristically though, the origin of the problem was tracked down and the necessary alterations made:

4 AUG. 1852.

12088.   The whole day almost in vain, for at the end of it I discovered a source of error which vitiated all the results and also those of yesterday—but it was well to know the error.   The curb (12068) which I used to confine the vibrations of the beam was of platina wire, the beam itself of glass, and these two when pressed together by torsion adhered with such force as not to allow the separation or displacement of the object when the magnetic force even surpassed the torsion force. No wonder the results yesterday were incomprehensible.

Searching always to make abstract ideas concrete, vocabulary was an ever present preoccupation. The 'iron house' referred to is nothing other than a Faraday cage; the phrase 'lines of force' remains current more than 140 years later:

10 SEPTR. 1854.

13341. Magnetic *lines of force* convey a far better and purer idea than the phrase magnetic current or magnetic flood: it avoids the assumption of a current or of two currents and also of fluids or a fluid, yet conveys a full and useful pictorial idea to the mind.

13 SEPTR. 1854.

13397. We live and experiment within a magnetic flood of force and subject to a mixed magnetic medium. We ought to live and experiment in a hard steel or an iron house. A good soft iron chamber will give a magnetic vacuum when it is wanted.

Delicate experiments with a galvanometer could be disturbed by small magnetic fields induced through surprising sources: but there was always an answer:

15766. Motion of my steel spectacles upon my head sadly disturbs the galvanometer needle. I must use the tortoise shell glasses.

A final attack on the relation between gravity and other forces, in early 1859, when Faraday was 68, shows his mind still actively probing for an experimental foothold:

10 FEBY. 1859.

15785. Surely the force of gravitation and its probable relation to other forms of force may be attacked by experiment. Let us try to think of some possibilities.

15786. Suppose a relation to exist between gravitation and electricity, and that as gravitation diminishes or increases by variation of distance, electricity either positive or negative were to appear—is not likely, nevertheless try, for less likely things apparently have happened in nature.

15787. There is more chance of any observable effect in a body acted on by the earth than in the same body acted on by a like body. There is more chance of a variation being observed in a ton of water or lead when lifted a hundred yards upwards from the earth, than in the same ton when removed a hundred yards in a horizontal direction from the side of another ton by which it at first stood.

The last entry of all in his laboratory notebook, on 12 March 1862, before he was finally forced into complete retirement through loss of memory, records a remarkable attempt to see whether the light emitted by inorganic salts in a flame changed its polarisation when the flame was in a magnetic field. In a similar experiment some 30 years later, using a spectroscope, the Dutch physicist Zeeman detected a splitting of the spectral lines under the conditions set up by Faraday, and was awarded the Nobel Prize for Physics in 1902 for his discovery:

12 March 1862.

Apparatus as on last day (28 Jany.) but only 10 pr. of Voltaic battery for the Electromagnet.

The colourless Gas flame ascended between the poles of the Magnet and the salts of Sodium, Lithium, etc. were used to give colour. A Nicol's polarizer was placed just before the intense magnetic field and an analyzer at the other extreme of the apparatus. Then the E. Magnet was made and unmade, but not the slightest trace of effect on or change of the lines in the spectrum were observed in any position of the polarizer or analyzer.

Chapter 9

Science in the Lecture Theatre

Faraday used his friend Benjamin Abbott (see Chapter 2) as a foil for working out some of his ideas and his correspondence as a means of focusing his mind on topics that seemed to him important. Having been attracted to the Royal Institution by Humphry Davy's lectures on chemistry, which had captured the public for many years, and learnt at close quarters about the art of lecturing with demonstrations through his job as Davy's assistant, Faraday already had decided opinions about it. In the following lengthy account he goes into detail about every aspect of the art, of which he became one of the greatest exponents. This account was spread over several letters to Abbott, written between 1 and 18 June 1813:

> Again I resort for pleasure (and to dispell the dullness of a violent headache) to my correspondence with you, 'tho perfectly unfit for it except as it may answer the purpose of amusing myself. The subject upon which I shall dwell more particularly at present has been in my head for some considerable time and now it bursts forth in all its confusion.
>
> In a word, Ben, I intend to give you my ideas on the subject of lectures and lecturers in general. The observations and ideas I shall set down are such as entered my mind at the moment the circumstances that gave rise to them took place. I shall point out but few beauties or few faults that I have

Faraday giving a Friday Evening Discourse, from *Illustrated London News* 1846.

not witnessed in the presence of a numerous assembly and it is exceedingly probable, or rather certain, that I should have noticed more of these particulars if I had seen more lecturers or, in other words, I do not pretend to give you an account of all the faults possible in a lecture or directions for the composing and delivering of a perfect one.

On going to a lecture I generally get there before it begins. Indeed, I consider it as an impropriety of no small magnitude to disturb the attention of an audience by entering amongst them in the midst of a lecture and, indeed, bordering upon an insult to the lecturer. By arriving there before the commencement, I have avoided this error and have had time to observe the lecture room.

The best form for a lecture room in general is without dispute a circular one, 'tho in particular circumstances deviations may with propriety be adopted. The seats should be arranged that no obstruction intervene between the spectator and the lecture table. If there is a gallery, each person in it should be situated in a manner the most convenient for observation and hearing. Those in which I have seen company and which please me most are the Theatre Royal, Haymarket, the Anatomical Theatre and the Lecture room here, to the last of which I must give preference. In a lecture room I would have the lecturer on a level with the lowest person in it. Consequently the pit should ascend very considerably, an object which cannot be attained in a theatre. Also, in the two I have mentioned, the lecturer is too far backward and a part of his audience is placed in a direction by far too oblique with respect to him (I allude to the side seats of this theatre).

In considering the form of a Lecture room we should take into account the time at which it is intended to be occupied, inasmuch as the modes of lighting by means natural and artificial are very different. In this particular the theatres in a large way have one advantage, i.e. in the site of their stage lamps which illuminate in a grand manner all before them, 'tho at the same time they fatigue the eyes of those who are situated low in the house. But 'tho Walker has shewn in the most splendid and sublime manner that Astronomy may be illustrated in a way the most striking by artificial light, yet, from what little I know of these things, I conceive that for by far the greater part of philosophy, day light is the most eligible and convenient.

When a Lecture room is illuminated by the light of the sun it should constantly be admitted (if convenient) at the top, not only as rendering the whole of the interior more uniform and distinct but also for the convenience of darkening the room. In the instance of the Lecture room here, you will readily recognize the mode of attaining that end to which I allude.

There is another circumstance to be considered with respect

to a Lecture room, of as much importance almost as light itself, and that is ventilation.   How often have I felt oppression in the highest degree when surrounded by a number of other persons and confined in one portion of air.   How have I wished the Lecture finished, the lights extinguished and myself away merely to obtain a fresh supply of that element. The want of it caused the want of attention and pleasure (and even of comfort) and not to be regained without its previous admission. Attention to this is more particularly necessary in a lecture room intended for night delivery, as the lights burning add considerably to the oppression produced on the body.

Entrance and Exit are things too, worthy of consideration amongst the particulars of a Lecture room, but I shall say no more on them than to refer you to the mode in which this is arranged here, a mode excellently well adapted for the convenience of a great number of persons.

Having thus thrown off in a cursory manner such thoughts as spontaneously entered my mind on this part of the subject it appears proper next to consider the subject fit for the purposes of a Lecture.   Science is undeniably the most eminent in its fitness for this purpose.   There is no part of it that may not be treated of, illustrated and explained with profit and pleasure to the hearers in this manner. The facility, too, with which it allows of manual and experimental illustrations place it foremost in this class of subject. After it comes (as I conceive) Arts and Manufactures, the Polite Arts, Belles Lettres etc, a list which may be extended until it includes almost every thought and idea in the mind of man (Politics excepted). I was going to add religion to the exception but remembered that is it explained and laid forth in the most popular and eminent manner in this way.

The fitness of subjects, however, is connected in an inseparable manner with the kind of audience that is to be present, since excellent Lectures in themselves would appear absurd if delivered before an audience that did not understand them. Anatomy would not do for the generality of audiences at the R.I., neither would Metaphysics engage the attention of a company of school boys. Let the subject fit the audience or otherwise success may be despaired of.

A Lecturer may consider his audience as being polite or vulgar (terms I wish you understand according to Shuffletons new Dictionary) learned or unlearned (with respect to the subject), listeners or gazers. Polite company expect to be entertained not only by the subject of the Lecture but by the manner of the Lecturer; they look for respect, for language consonant to their dignity and ideas on a level with their own. The vulgar (that is to say, in general, those who will take the trouble of thinking) and the bees of business wish for something that they can comprehend. This may be deep and elaborate for the learned but, for those who are as yet tyros and unacquainted with the subject, must be simple and plain. Lastly, listeners expect reason and sense, whilst gazers only require a succession of words.

These considerations should all of them engage the attention of the Lecturer whilst preparing for his occupation, each particular having an influence on his arrangements proportionate to the nature of the company he expects. He should consider them connectedly, so as to keep engaged completely during the whole of the Lecture the attention of his audience.

The hour at which a Lecture should be delivered should be considered at the same time with the nature of the audience we expect or wish for. If we would suit a particular class of persons we must fix it at the hour most convenient for them. If we would wish to exclude any, let the time be such that they cannot attend it. In general we may distinguish them according to their times into morning and evening Lectures, each being adapted for different classes of persons.

I need not point out to the active mind of my friend the astonishing disproportion (or rather difference) in the perceptive powers of the eye and the ear and the facility and clearness with which the first of these organs conveys ideas to the mind; ideas which, being thus gained, are held far more retentively and firmly in the memory than when introduced by the ear. Tis true the ear here labours under a disadvantage, which is, that the Lecturer may not always be qualified to state a fact with the utmost precision and clearness that language allows him and that the ear can understand, and thus the complete action of the organ (or rather of its assigned portion of the sensorium) is not called forth.   But this evidently points out to us the necessity of aiding it by using the eye also as a medium for the attainment of knowledge, and strikingly shews the necessity of apparatus.

Apparatus, therefore, is an essential part of every lecture in which it can be introduced, but to apparatus should be added at every convenient opportunity illustrations that may not perhaps deserve the name of apparatus and of experiments, and yet may by introduced with considerable force and effect in proper places.   Diagrams & Tables, too, are necessary, or at least add in an eminent degree to the illustration and perfection of a Lecture.

When an experimental Lecture is to be delivered and apparatus is to be exhibited, some kind of order should be observed in the arrangement of them on the Lecture table. Every particular part illustrative of the Lecture should be in view; no one thing should hide another from the audience, nor should anything stand in the way of, or obstruct, the Lecturer. They should be so placed, too, as to produce a kind of uniformity in appearance.   No one part should appear naked and another crowded, unless some particular reason exists and makes it necessary to be so.   At the same time the whole should be so arranged as to keep one operation from interfering with another.   If the Lecture table appears crowded, if the Lecturer (hid by the apparatus) is invisible, if things appear crooked or aside or unequal, or if some are out of sight (and this without any particular reason), the Lecturer is considered (and with reason too) as an awkward contriver and a bungler.

Diagrams ('tho ever so rough) are often times of important use in a lecture. The facility with which they illustrate ideas, and the diversity they produce in the circumstances occurrant, render them highly agreeable to an audience. By diagrams, I do not mean drawings (nor do I exclude drawings) but a plain and simple statement in a few lines of what requires many words. A sheet of cartridge paper and a pen, or a black board and chalk, are often times of great importance. I, in general, allude to temporary diagrams and would resort to temporary means to obtain them.

A diagram or a table (by which I mean constituent parts or proportions wrote out in a rough enlarged way) should be left in view of the audience for a short time after the Lecturer himself has explained them, that they may arrange the ideas contained in them in their minds and also refer to them in any other parts of the theory connected with the same subject and (if they choose, as is often the case) also to copy them.

With respect to illustrations (simply so called) no regular rules can be given on them. They must be in part extempore and suggested to the mind of the Lecturer by particular circumstances; they may be at one time proper at another time improper, but they should always be striking and to the point.

The most prominent requisite to a Lecturer, 'tho perhaps not really the most important, is a good delivery. For though, to all true philosophers, Science and Nature will have charms innumerable in every dress, yet I am sorry to say that the generality of mankind cannot accompany us one short hour unless the path is strewn with flowers. In order, therefore, to gain the attention of an audience (and what can be more disagreeable to a Lecturer than the want of it) it is necessary to pay some attention to the manner of expression. The utterance should not be rapid and hurried, and consequently unintelligible, but slow and deliberate, conveying ideas with ease from the Lecturer and infusing them with clearness and readiness into the minds of the audience. A Lecturer should endeavour by all means to obtain a facility of utterance and the power of clothing his thoughts and ideas in language smooth and harmonious and, at the same time, simple and easy. His periods should be round, not too long or unequal; they should be complete and expressive, conveying clearly the whole of the ideas intended to be conveyed. If they are long or obscure or incomplete they give rise to a degree of labour in the minds of the hearers which quickly causes lassitude, indifference and even disgust.

With respect to the action of a Lecturer, it is requisite that he should have some, 'tho it does not here bear the importance that it does in other branches of oratory. For 'tho I know of no species of delivery (divinity excepted) that requires less motion, yet I would by no means have a Lecturer glued to the table or screwed on the floor. He must by all means appear as a body distinct and separate from the things around him, and must have some motion apart from that which they possess.

A Lecturer should appear easy & collected, undaunted & unconcerned, his thoughts about him and his mind clear and free for the contemplation and description of his subject. His action should not be hasty and violent but slow, easy and natural, consisting principally in changes of the posture of the body in order to avoid the air of stiffness or sameness that would otherwise be unavoidable. His whole behaviour should evince respect for his audience and he should in no case forget that he is in their presence. No accident that does not interfere with their convenience should disturb his serenity, or cause variation in his behaviour. He should never (if possible) turn his back on them, but should give them full reason to believe that all his powers have been exerted for their pleasure and instruction.

Some Lecturers choose to express their thoughts extemporaneously, immediately as they occur to the mind, whilst others previously arrange them and draw them forth on paper. Those who are of the first description are certainly more unengaged and more at liberty to attend to other points of delivery than their pages; but as every person on whom the duty falls is not equally competent for the prompt clothing and utterance of his matter, it becomes necessary that the second method should be resorted to. This mode, too, has its advantages, inasmuch as more time is allowed for the arrangement of the subject and more attention can be paid to the neatness of expression.

But, 'tho I allow a Lecturer to write out his matter, I do not approve of his reading it, at least not as he would a quotation or extract. He should deliver it in a ready and free manner, referring to his book merely as he would to copious notes, and not confining his tongue to the exact path there delineated, but digress as circumstances may demand, or localities allow.

A Lecturer should exert his utmost efforts to gain completely the mind and attention of his audience and irresistibly to make them join in his ideas to the end of the subject. He should endeavour to raise their interest at the commencement of the Lecture and, by a series of imperceptible gradations unknown by the company, keep it alive as long as the subject demands it. No breaks or digressions foreign to the purpose should have a place in the circumstances of the evening; no opportunity should be allowed to the audience in which their minds could wander from the subject or return to inattention and carelessness. A flame should be ignited at the commencement and kept alive with unremitting splendour to the end. For this reason I very much disapprove of breaks in a Lecture and, where they can by any means be avoided, they should on no account find place. If it is unavoidably necessary (to complete the arrangement of some experiment, or for other reasons) to leave some experiments in a state of progression, state some peculiar circumstance to employ as much as possible the minds of the audience during the unoccupied space: but if possible avoid it.

Digressions and wanderings produce more or less the bad effects of a complete break or delay in a Lecture, and should therefore never be. Surely such an error in the character of a Lecturer cannot require pointing out, even to those who resort to it; its impropriety must be evident and I should perhaps have done well to pass it.

Before, however, I quite leave this part of my subject I would wish to notice a point in some manner connected with it. In Lectures (and more particularly some experimental ones) it will at times happen that accidents or other incommoding circumstances take place. On these occasions an apology is sometimes necessary, but not always. I would wish apologies to be made as seldom as possible, and generally only when the inconvenience extends to the company's. I have several times seen the attention of by far the greater part of the audience called to an error by the apology that followed it.

An experimental Lecturer should attend very carefully to the choice he may make of experiments for the illustration of his subjects. They should be important as they respect the science they are applied to, yet clear and such as may easily and generally be understood. They should rather approach to simplicity, and explain the established principles of the subject, than be elaborate and apply to minute phenomena only. I speak here (be it understood) of those lectures which are delivered before a mixed audience, and the nature of which will not admit of their being applied to the explanation of any but the principle parts of a science. If, to a particular audience, you dwell on a particular subject, still adhere to the same precept, 'tho perhaps not exactly to the same rule. Let your experiments apply to the subject you elucidate. Do not introduce those that are not to the point.

'Tho this last part of my letter may appear superfluous, seeing that the principle is so evident to every capacity, yet

I assure you, dear Abbott, I have seen it broken through in the most violent manner. A mere ale-house trick has more than once been introduced in a Lecture delivered not far from Pall-Mall as an elucidation of the laws of motion.

Neither should too much stress be laid upon what I would call small experiments, or rather illustrations. It pleases me well to observe a neat idea enter the head of a Lecturer, the which he will immediately and aptly illustrate or explain by a few motions of his hand, a card, a lamp, a glass of water or any other thing that may be by him. When he calls your attention in a particular way to a decisive experiment that has entered his mind, clear and important in its application to the subject and then—lets fall a card, I turn with disgust from the Lecturer and his experiments.

'Tis well, too, when a Lecturer has the ready wit and the presence of mind to turn any casual circumstance to an illustration of his subject. Any particular circumstance that has become table talk for the town, any local advantages or disadvantages, any trivial circumstance that may arise in company, give great force to illustrations aptly drawn from them, and please the audience highly, as they conceive they perfectly understand them.

Apt experiments (to which I have before referred) ought to be explained by satisfactory theory, or otherwise we merely patch an old coat with new cloth and the whole (hole) becomes worse. If a satisfactory theory can be given, it ought to be given. If we doubt a received opinion, let us not leave it unnoticed and affirm our own ideas, but state it clearly and lay down also our objections. If the scientific world is divided in opinion, state both sides of the question) and let each one judge for himself, assisting him by noticing the most striking and forcible circumstances on each side. Then (and then only) shall we do justice to the subject, please the audience and satisfy our honour (the honour of a philosopher).

Chapter 10

Science for Young People

Though he delighted in lecturing to adults, especially in Friday Evening Discourses at the Royal Institution, the apotheosis of Faraday's lecturing skill lay in his lectures for young people. The series of lecture-demonstrations, as he put it 'for a juvenile auditory', started at Christmas time in 1826 and, apart from a brief interlude during the Second World War, continue to this day. In recent years, through the medium of television, what became known as the Christmas lectures reach audiences averaging more that one and a half million for each lecture. Despite the presence of the television cameras, the core spirit of the lectures remains as it always was: eminent practising scientists describe the essence of their work in simple language, to 450 teenagers in the lecture theatre at the Royal Institution, using the rostrum where Faraday himself spoke. As in Faraday's time the subject is brought alive by many illustrations, and in particular by demonstrations.

The Chemical History of a Candle

Faraday did not give the first series of Christmas lectures himself, but he did give the second, simply entitled 'Chemistry', and thereafter no fewer that sixteen more, up till 1860. Easily the most famous are the ones on 'The Chemical History of a Candle', which were so successful that they were repeated several times, as the excerpt below makes clear. Faraday's mode of exposition was masterly: wishing to take his young audience into some of the most contemporary areas of chemical science by the gentlest of pathways, he used an ordinary candle as the focus for his presentation. When the lectures in question were first given there could have been no more commonplace object: every dinner table had one on it. Yet such was Faraday's genius, that he was able to examine every aspect of the candle and its flame, so as to induce wonder at the marvels of understanding which, even in the 1850s, lay behind the simple facts. What is a candle made of? How does it burn? Why does its flame have such a shape? What lies in the yellow periphery? Or the transparent centre? How is the heat generated? Or transported? One could go on.

Although the Christmas lectures on the topic were given first in 1848, the resulting book called *The Chemical History of a Candle* was first published in 1861. It has scarcely been out of print ever since. Today the largest sales are in Japan, where it is given as a set book to school children to learn how observational science works. In a book like the present one, which aims to present Faraday's own words, it has to be said that the lectures were actually taken down and transcribed by others for publication. Still, the essence of the man and his approach shines through. The full text runs to 150 pages so only very brief glimpses are possible here. The opening sets the scene:

> I purpose to bring before you, in the course of these lectures, the Chemical History of a Candle. I have taken this subject on a former occasion, and were it left to my own will I should prefer to repeat it almost every year; so abundant is the interest that attaches itself to the subject, so wonderful are the varieties of outlet which it offers into the various departments of philosophy. There is not a law under which any part of this universe is governed which does not come into play and is touched upon in these phenomena. There is no better, there is no more open door by which you can enter into the study of natural philosophy, than by considering the physical phenomena of a candle.

And before proceeding, let me say this also: that though our subject be so great, and our intention of treating it honestly, seriously and philosophically, yet I mean to pass away from all those who are seniors among us. I claim the privilege of speaking to juveniles as a juvenile myself. I have done so on former occasions and, if you please, I shall do so again. And though I stand here with the knowledge of having the words I utter given to the world, yet that shall not deter me from speaking in the same familiar way to those whom I esteem nearest to me on this occasion.

And now, my boys and girls, I must first tell you of what candles are made. Some are great curiosities. I have here some bits of timber, branches to trees particularly famous for their burning. And here you see a piece of that very curious substance taken out of some of the bogs in Ireland, called *candle-wood*; a hard, strong, excellent wood, evidently fitted for good work as a resister of force, and yet withal burning so well that where it is found they make splinters of it, and torches, since it burns like a candle, and gives a very good light indeed. And in this wood we have one of the most beautiful illustrations of the general nature of a candle that I can possibly give. The fuel provided, the means of bringing that fuel to the place of chemical action, the regular and gradual supply of air to that place of action—heat and light—all produced by a little piece of wood of this kind, forming, in fact, a natural candle.

But we must speak of candles as they are in commerce. Here are a couple of candles commonly called dips. They are made of lengths of cotton cut off, hung up by a loop, dipped into melted tallow, taken out again and cooled, then re-dipped, until there is an accumulation of tallow round the cotton. In order that you may have an idea of the various characters of these candles, you see these which I hold in my hand—they are very small and very curious. They are, or were, the candles used by the miners in coal-mines. In olden times the miner had to find his own candles, and it

The candle flame.

was supposed that a small candle would not so soon set fire
to the fire-damp in the coal-mines as a large one; and for
that reason, as well as for economy's sake, he had candles
made of the sort—20, 30, 40, or 60 to the pound. They have
been replaced since then by the steel-mill, and then by the
Davy-lamp, and other safety-lamps of various kinds. I have
here a candle that was taken out of the *Royal George*, it is
said, by Colonel Pasley. It has been sunk in the sea for
many years, subject to the action of salt water. It shows
you how well candles may be preserved; for though it is
cracked about and broken a good deal, yet when lighted it
goes on.

Next, he considers how the light is produced, how the fuel
gets to the flame, from the solid parts of the candle, through the
liquid phase at the base of the flame, up the wick and into the
flame itself. Imagine that all the while he is talking (for this is the
transcript of speech) he is manipulating, teasing the candle and its
flame to bring out the causes within:

Now as to the light of the candle. We will light one or two, and set them at work in the performance of their proper functions. You observe a candle is a very different thing from a lamp. With a lamp you take a little oil, fill your vessel, put in a little moss or some cotton prepared by artificial means, and then light the top of the wick. When the flame runs down the cotton to the oil, it gets extinguished, but it goes on burning in the part above. Now, I have no doubt, you will ask, how is it that the oil which will not burn of itself gets up to the top of the cotton where it will burn? We shall presently examine that; but there is a much more wonderful thing about the burning of a candle than this. You have here a solid substance with no vessel to contain it; and how is it that this solid substance can get up to the place where the flame is? How is it that this solid gets there, it not being a fluid? Or, when it is made a fluid, then how is it that it keeps together? This is a wonderful thing about a candle.

The vagaries of experiment, such as are caused by the wind in the lecture theatre, are turned to good account in the exposition and (with some homely illustrations from the street) show how the melted wax gets to the flame:

We have here a good deal of wind, which will help us in some of our illustrations, but tease us in others. For the sake, therefore, of a little regularity, and to simplify the matter, I shall make a quiet flame. For who can study a subject when there are difficulties in the way not belonging to it? Here is a clever invention of some costermonger or street-stander in the market-place for the shading of their candles on Saturday nights, when they are selling their greens, or potatoes, or fish. I have very often admired it. They put a lamp-glass round the candle, supported on a kind of gallery, which clasps it, and it can be slipped up and down as required. By the use of this lamp-glass, employed in the same way, you have a steady flame which you can look at, and carefully examine (as I hope you will do) at home.

You see then, in the first instance, that a beautiful cup is formed. As the air comes to the candle it moves upwards by the force of the current which the heat of the candle produces, and it so cools all the sides of the wax, tallow, or fuel, as to keep the edge much cooler than the part within. The part within melts by the flame that runs down the wick as far as it can go before it is extinguished, but the part on the outside does not melt. If I made a current in one direction, my cup would be lop-sided, and the fluid would consequently run over, for the same force of gravity which holds worlds together holds this fluid in a horizontal position, and if the cup be not horizontal, of course the fluid will run away in a guttering. You see, therefore, that the cup is formed by this beautifully regular ascending current of air playing upon all sides, which keeps the exterior of the candle cool.

General conclusions (even with a moral tinge?) are drawn from the existence of imperfection in man-made objects, but the young listeners were left with a powerful message, not just about science but about utility and beauty:

You see now why you would have had such a bad result if you were to burn these beautiful candles that I have shown you, which are irregular, intermittent in their shape, and cannot, therefore, have that nicely formed edge to the cup which is the great beauty in a candle. I hope you will now see that the perfection of a process (that is, its utility) is the better point of beauty about it. It is not the best looking thing, but the best acting thing, which is the most advantageous to us.

Now the greatest mistakes and faults with regards to candles, as in many other things, often bring with them instruction which we should not receive if they had not occurred. We come here to be philosophers, and I hope you will always remember that whenever a result happens, especially if it be new, you should say, 'what is the cause? Why does it occur?' and you will in the course of time find out the reason.

Next, how does the fuel get up to the flame? A homely analogy brings the principle of capillary attraction to life:

But how does the flame get hold of the fuel? There is a beautiful point about that—*capillary attraction*. 'Capillary attraction?' you say—'the attraction of hairs.' Well, never mind the name, it was given in old times before we had a good understanding of what the real power was. It is by what is called capillary attraction that the fuel is conveyed to the part where combustion goes on, and is deposited there, not in a careless way, but very beautifully in the very midst of the centre of action, which takes place around it. Now I am going to give you one or two instances of capillary attraction. It is that kind of action or attraction which makes two things that do not dissolve in each other still hold together. When you wash your hands, you wet them thoroughly; you take a little soap to make the adhesion better, and you find your hand remains wet. This is by that kind of attraction of which I am about to speak. And what is more, if your hands are not soiled (as they almost always are by the usages of life), if you put your finger into a little warm water, the water will creep a little way up the finger, though you may not stop to examine it.

When you wash your hands you take a towel to wipe off the water, and it is by that kind of wetting, or that kind of attraction which makes the towel become wet with water, that the wick is made wet with the tallow. I have known some careless boys and girls (indeed, I have known it happen to careful people as well) who, having washed their hands and wiped them with a towel, have thrown the towel over the side of the basin, and before long it has drawn all the water out of the basin and conveyed it to the floor, because it happened to be thrown over the side in such a way as to serve the purpose of a syphon.

After discovering in detail the process of combustion, and how the carbon in the candle is turned into carbon dioxide by combining with oxygen in the atmosphere, Faraday turns to the wider question of how human beings derive their energy from food, also through its reaction with oxygen. But first we must take in the oxygen from the air, which we do by respiration. The analogy with the candle is beautifully developed:

Let us now go a little further. What is all this process going on within us which we cannot do without, either day or night, which is so provided for by the Author of all things that He has arranged that it shall be independent of all will? If we restrain our respiration, as we can to a certain extent, we should destroy ourselves. When we are asleep, the organs of respiration and the parts that are associated with them, still go on with their action, so necessary is this process of respiration to us, this contact of the air with the lungs. I must tell you, in the briefest possible manner, what this process is. We consume food: the food goes through that strange set of vessels and organs within us, and is brought into various parts of the system, into the digestive parts especially; and alternately the portion which is so changed is carried through our lungs by one set of vessels, while the air that we inhale and exhale is drawn into and thrown out of the lungs by another set of vessels, so that the air and the food come close together, separated only by an exceedingly thin surface: the air can thus act upon the blood by this process, producing precisely the same results in kind as we have seen in the case of the candle. The candle combines with parts of the air, forming carbonic acid, and evolves heat; so in the lungs there is this curious, wonderful change taking place. The air entering, combines with the carbon (not carbon in a free state, but, as in this case, placed ready for action at the moment), and makes carbonic acid, and is so thrown out into the atmosphere, and thus this singular result takes place. We may thus look upon the food as fuel.

Candles burn in air, and we 'burn' food to supply us with our energy, in both cases producing carbon dioxide, yet in plants the opposite appears to be the case: they absorb carbon dioxide and turn it into carbohydrates by the process that nowadays we call photosynthesis. The bowl of goldfish and the sprigs of green plants were vivid illustrations of the point:

> As charcoal burns it becomes a vapour and passes off into the atmosphere, which is the great vehicle, the great carrier for conveying it away to other places. Then what becomes of it? Wonderful is it to find that the change produced by respiration, which seems so injurious to us (for we cannot breathe air twice over) is the very life and support of plants and vegetables that grow upon the surface of the earth. It is the same also under the surface, in the great bodies of water, for fishes and other animals respire upon the same principle, though not exactly by contact with the open air.

> Such fish as I have here [pointing to a globe of gold-fish] respire by the oxygen which is dissolved from the air by the water, and from carbonic acid, and they all move about to produce the one great work of making the animal and vegetable kingdoms subservient to each other.

> And all the plants growing upon the surface of the earth, like that which I have brought here to serve as an illustration, absorb carbon. These leaves are taking up their carbon from the atmosphere to which we have given it in the form of carbonic acid, and they are growing and prospering. Give them a pure air like ours, and they could not live in it; give them carbon with other matters, and they live and rejoice. This piece of wood gets all its carbon, as the trees and plants get theirs, from the atmosphere, which, as we have seen, carries away what is bad for us and at the same time good for them: what is disease to the one being health to the other. So are we made dependent, not merely upon our fellow-creatures, but upon our fellow-existers, all Nature being tied together by the laws that make one part conduce to the good of another.

Both in the candle and in the human body, the paradox of

combustion is that it does not take place spontaneously. The candle, and the coal gas used in the demonstration, must be ignited before they burn, though a hot taper applied to gunpowder produced a predictable (and no doubt satisfyingly explosive) reaction. But still the analogy with human metabolism is brought back: the body appears to burn its fuel at room temperature. Today we know that catalysts called enzymes make this apparently improbable process happen:

> It is a striking thing to see that the matter which is appointed to serve the purpose of fuel *waits* in its action; it does not start off burning, like the lead and many other things that I could show you, but which I have not encumbered the table with, but it waits for action. This waiting is a curious and wonderful thing. Candles—those Japanese candles, for instance—do not start into action at once like the lead or iron (for iron finely divided does the same thing as lead), but there they wait for years, perhaps for ages, without undergoing any alteration. I have here a supply of coal-gas. The jet is giving forth the gas, but you see it does not take fire. It comes out into the air, but it waits till it is hot enough before it burns. If I make it hot enough, it takes fire. If I blow it out, the gas that is issuing forth waits till the light is applied to it again.

> It is curious to see how different substances wait—how some will wait till the temperature is raised in a little, and others till it is raised a good deal. I have here a little gun-powder and some gun-cotton; even these things differ in the conditions under which they will burn. The gunpowder is composed of carbon and other substances, making it highly combustible, and the gun-cotton is another combustible preparation. They are both waiting, but they will start into activity at different degrees or heat, or under different conditions. By applying a heated wire to them, we shall see which will start first [touching the gun-cotton with the hot iron]. You see the gun-cotton has gone off, but not even the hottest part of the wire is now hot enough to fire the gunpowder. How beautifully

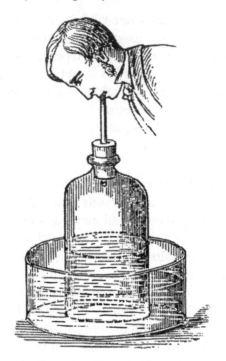

Breathing out carbon dioxide.

that shows you the difference in the degree in which bodies act in this way! In the one case the substance will wait any time until the associated bodies are made active by heat, but in the other, as in the process of respiration, it waits no time.

In the lungs, as soon as the air enters, it unites with the carbon. Even in the lowest temperature which the body can bear, short of being frozen, the action begins at once, producing the carbonic acid of respiration; and so all things go on fitly and properly. Thus you see the analogy between respiration, and combustion is rendered still more beautiful and striking.

Faraday's envoi to his young listeners mirrors the sentences with which sermons are traditionally closed: 'Let your light so shine before men that they may see your good works, and glorify

your Father which is in Heaven'.   Faraday must have heard
the biblical phrase many times in his Sunday chapel devotions.
Indeed, he was twice an elder in the Sandemanian Church and
frequently preached sermons. However, he chose a candle as the
centrepiece for his lectures, not for religious reasons but because it
epitomised the science that he wanted to expound. Nevertheless,
its symbolism did not escape him:

> Indeed, all I can say to you at the end of these lectures (for
> we must come to an end at one time or another) is to express
> a wish that you may, in your generation, be fit to compare to
> a candle; that you may, like it, shine as lights to those about
> you; that, in all your actions, you may justify the beauty of
> the taper by making your deeds honourable and effectual in
> the discharge of your duty to your fellow-men.

On the Various Forces of Nature

Faraday's purpose in the *Chemical History of a Candle* was (as the
title implies) to introduce his young audience to those elements
central to chemistry, as it was known at the time.   These are
the composition of substances and nature of chemical reactions
that transform one into another.   However, as we have seen
from his correspondence, and his own musings in his laboratory
notebook, for much of his career Faraday's preoccupations led him
far beyond chemistry. The drive that dominated his later work was
the search for connections between the numerous natural forces
that, at first sight, appeared unrelated but, with deeper probing
turned out to have unexpected connections. Thus chemical energy
can produce electricity through the action of a battery, an electrical
current passing through a wire gives rise to a magnetic field, a
magnetic field rotates the plane of polarisation of a light beam and
so on. Faraday even tried unsuccessfully to connect magnetism
with gravity (an effort that continues to this day). It is against this
background that the second series of Christmas lectures given by
Faraday, and published in book form, was presented.

In contrast to the chemical focus of the lectures on the candle, the aim of the ones entitled 'On the Various Forces of Nature and their Relations to Each Other' was to introduce his young audience to the elements of physics, as they were known in 1860. By comparison with the *Chemical History of a Candle*, which has remained perennially popular up to the present day, the *Various Forces of Nature* was reprinted only a few times and the latest edition is dated 1894. Like the book on the candle, it was not written by Faraday himself, but was compiled from notes taken as he spoke. The text was then edited for publication by Sir William Crookes, himself a distinguished scientist whose name was given to the vacuum discharge tube. We can be sure, nevertheless, that the text adheres closely to what the lecturer actually said since, as Crookes writes in his introduction: 'the lectures are printed as they were spoken, *verbatim et literatim*. A careful and skilful reporter took them down, and the manuscript, as deciphered from his notes, was subsequently most carefully corrected by the editor'.

Faraday opens his first lecture with a clear statement of his aims: to talk to the young (he himself was 69). His evocation of wonder at the simplest and most obvious facts of nature is among the most eloquent among all his writings. It leads quite logically into several graphic illustrations of the first of the natural forces that he wished to consider, namely gravity:

> I shall here claim, as I always have done on these occasions, the right of addressing myself to the younger members of the audience. And for this purpose, therefore, unfitted as it may seem for an elderly infirm man to do so, I will return to second childhood and become, as it were, young again amongst the young.

> Let us now consider for a little while, how wonderfully we stand upon this world. Here it is we are born, bred, and live, and yet we view these things with an almost entire absence of wonder to ourselves respecting the way in which all this happens. So small, indeed, is our wonder, that we are never taken by surprise; and I do think that, to a young

person of ten, fifteen, or twenty years of age, perhaps the first sight of a cataract or a mountain would occasion him more surprise that he had ever felt concerning the means of his own existence,—how he came here; how he lives; by what means he stands upright; and through what means he moves about from place to place. Hence, we come into this world, we live, and depart from it, without our thoughts being called specifically to consider how all this takes place. And were it not for the exertions of some few inquiring minds, who have looked into these things and ascertained the very beautiful laws and conditions by which we *do* live and stand upon the earth, we should hardly be aware that there was anything wonderful in it.

These inquiries, which have occupied philosophers from the earliest days, when they first began to find out the laws by which we grow, and exist and enjoy ourselves, up to the present time, have shewn us that all this was effected in consequence of the existence of certain forces, or *abilities* to do things, or *powers*, that are so common that nothing can be more so. For nothing is commoner than the wonderful powers by which we are enabled to stand upright. They are essential to our existence every moment.

We are not to suppose that there are so very many different powers. On the contrary, it is wonderful to think how few are the powers by which all the phenomena of nature are governed. And thus, beginning with the simplest experiments of pushing and pulling, I shall gradually proceed to distinguish these powers one from the other, and compare the way in which they combine together. This world upon which we stand (and we have not much need to travel out of the world for illustrations of our subject; but the mind of man is not confined like the matter of his body, and thus he may and does travel outwards; for wherever his sight can pierce, there his observations can penetrate) is pretty nearly a round globe, having its surface disposed in a manner of which this terrestrial globe by my side is a rough model. So much is land and so much is water, and by looking at it here we see in a sort of map or picture how the world is formed upon its surface.

When we come to dig into or examine it (as man does for his own instruction and advantage, in a variety of ways), we see that it is made up of different kinds of matter, subject to a very few powers, and all disposed in this strange and wonderful way, which gives to man a history—and such a history—as to what there is in those veins, in those rocks, the ores, the water springs, the atmosphere around, and all varieties of material substances, held together by means of *forces* in one great mass, 8,000 miles in diameter, that the mind is overwhelmed in contemplation of the wonderful history related by these strata (some of which are fine and thin like sheets of paper), all formed in succession by the forces of which I have spoken.

After talking about gravity, Faraday goes on to a quite different force of attraction, magnetism. The following extract gives a good impression of the way that he used demonstrations to make physical phenomena come alive:

All bodies attract each other at sensible distances. Suppose I take a few iron particles [dropping some small fragments of iron on the table]. There, I have already told you that in all cases where bodies fall, it is the *particles* that are attracted. You may consider these then as separate particles magnified, so as to be evident to your sight; they are loose from each other.

Here I have an arch of iron filings regularly built up like an iron bridge, because I have put them within a sphere of action which will cause them to attract each other. See!—I could let a mouse run through it, and yet if I try to do the same thing with them *here* [on the table], they do not attract each other at all. It is *that* [the magnet] which makes them hold together in the form of an elliptical bridge, so do the different particles of iron which constitute this nail hold together and make it

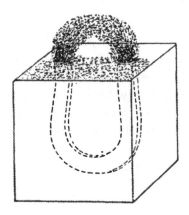

Arch made of iron filings.

one. And here is a bar of iron—why, it is only because the different parts of *this* iron are so wrought as to keep close together by the attraction *between* the particles that it is held together in one mass.

Attraction between the particles of which a bulk substance is composed may take place through more than one agency. The demonstration described in the previous extract showed in a very graphic way how magnetism brought a lot of iron filings together into a single lump. Now he shows how, merely by pressing them together, pieces of a soft material like lead can be made to stick. A boy's experiment was always the best for him; the 'force' he is describing is what we would call cohesion:

How can we make this attraction of the particles a little more simple? There are many things which, if brought together properly, will show this attraction. Here is a boy's experiment (and I like a boy's experiment). Get a tobacco-pipe, fill it with lead, melt it, and then pour it out upon a stone, and thus get a clean piece of lead (this is a better plan than scraping it—scraping alters the condition of the surface of the lead). I have here some pieces of lead which I melted this morning for the sake of making them clean. Now these pieces of lead hang together by the attraction of their particles, and if I press these two separate pieces close together, so as to bring their particles within the sphere of attraction, you will see how soon they become one. I have merely to give them a good squeeze, and draw the upper piece slightly round at the same time, and here they are as one, and all the bending and twisting I can give them will not separate them again. I have joined the lead together, not with solder, but simply by means of the attraction of the particles.

It is the varying degrees of cohesion between the particles that leads to the difference between solids, liquids and gases, exemplified by ice, liquid water and steam. In this case it is heat that causes the transition between the three states of matter. When it hit the block of ice, the red hot iron ball must have made a splendidly spectacular cloud of steam. And yet again we find the homely example face to face with the grand perspective—from arctic to tropics:

And now we have another point to examine, and this water is again a very good substance to take as an illustration (as philosophers we call it all water, even though it be in the form of ice or steam). Why is this water hard? [pointing to a block of ice] Because the attraction of the particles to each other is sufficient to make them retain their places in opposition to force applied to it. But what happens when we make the ice warm? Why, in that case we diminish to such a large extent the power of attraction that the solid substance is destroyed altogether. Let me illustrate this: I will take a red-hot ball of iron [Mr Anderson, by means of a pair of tongs, handed to the Lecturer a red-hot ball of iron, about two inches in diameter], because it will serve as a convenient source of heat [placing the red-hot iron in the centre of the block of ice]. You see I am now melting the ice where the iron touches it. You see the iron sinking into it, and while part of the solid water is becoming liquid, the heat of the ball is rapidly going off. A certain part of the water is actually rising in steam—the attraction of some of the particles is so much diminished that they cannot even hold together in the liquid form, but escape as vapour. At the same time, you see I cannot melt all this ice by the heat contained in this ball. In the course of a very short time I shall find it will have become quite cold.

Is it not a glorious thing for us to look at the sea, the rivers, and so forth, and to know that this same body in the northern regions is all solid ice and icebergs, while here, in a warmer climate, it has its attraction of cohesion so much diminished as to be liquid water? I remember once, when I was a boy, hearing of a trick in the country alehouse; the point was how to melt ice in a quart-pot by the fire, and freeze it to the stool. Well, the way they did it was this: they put some pounded ice in a pewter pot and added some salt to it, and the consequence was, that when the salt was mixed with it, the ice in the pot melted (they did not tell me anything about the salt) and they set the pot by the fire, just to make the result more mysterious. And in a short time the pot and the stool were frozen together, as we shall very shortly find it to be the case here. And all because salt has the power of lessening the attraction between the particles of ice. Here you see the tin dish is frozen to the board. I can even lift this little stool up by it.

Having introduced heat as a way of influencing the degree of cohesion between particles (or, as we would say, atoms and molecules), Faraday also connects the production of heat with chemical reaction, and by other means, too, like mechanical friction:

You know, as a matter of fact, no doubt, that when bodies burn they give out heat, but it is a curious thing that this heat does not continue. The heat goes away as soon as the action stops, and you see thereby that it depends upon the action *during the time* it is going on. It is not so with gravitation: this force is continuous, and is just as effective in making that lead press on the table as it was when it first fell there. Nothing occurs there which disappears when the action of falling is over. The pressure is upon the table, and will remain there until the lead is removed, whereas, in the action of chemical affinity to give light and heat, they go away immediately the action is over.

This lamp *seems* to evolve heat and light continuously, but it is owing to a constant stream of air coming into it on all sides, and this work of producing light and heat by chemical affinity will subside as soon as the stream of air is interrupted.

What is heat? We recognise heat by its power of liquefying solid bodies and vaporising liquid bodies, by its power of setting in action (and very often overcoming) chemical affinity. Think, how do we obtain heat? We obtain it in various ways, most abundantly by means of the chemical affinity we have just before been speaking about, but we can also obtain it in many other ways.

Friction will produce heat. The Indians rub pieces of wood together until they make them hot enough to take fire, and such things have been known as two branches of a tree rubbing together so hard as to set the tree on fire. I do not suppose I shall set these two pieces of wood on fire by friction, but I can readily produce heat enough to ignite some phosphorus.

In later lectures in the series, the forces that Faraday considers become more subtle and profound, as are the connections between them. The first is electrical attraction, exemplified by the static electric charge that can be generated by rubbing an amber rod with a cloth. The point is nicely made that before it is endowed with electric charge it is already subject to the other forces of gravitation, cohesion and chemical affinity dealt with in earlier lectures:

I wonder whether we shall be too deep today or not. Remember that we spoke of the attraction by gravitation of *all* bodies to all bodies by their simple approach. Remember that we spoke of the attraction of particles of the *same* kind to each other, the power which keeps them together in masses (iron attracted to iron, brass to brass, or water to water). Remember that we found, on looking into water, that there were particles of two different kinds attracted to each other, and this was a great step beyond the first simple attraction of gravitation because here we deal with attraction between *different* kinds of matter.

To-day we come to a kind of attraction even more curious than the last, namely, the attraction which we find to be of a double nature, of a curious and dual nature. And I want first of all to make the nature of this doubleness clear to you. Bodies are sometimes endowed with a wonderful attraction, which is not found in them in their ordinary state. For instance, here is a piece of shellac, having the attraction of gravitation, having the attraction of cohesion and, if I set fire to it, it would have the attraction of chemical affinity to the oxygen in the atmosphere. Now, all these powers we find *in* it as if they were parts of its substance, but there is another property which I will try and make evident by means of this ball, this bubble of air [a light india-rubber ball, inflated and suspended by a thread]. There is no attraction between this ball and this shellac present: there may be a little wind in the room slightly moving the ball about, but there is no attraction. But if I rub the shellac with a piece of flannel [rubbing the shellac, and then holding it near the ball], look at the attraction which has arisen out of the shellac simply by this friction, and which I may take away as easily by drawing it gently through my hand.

This, then, is sufficient in the outset to give you an idea of the nature of the force which we call ELECTRICITY.

Finally Faraday comes to magnetism. Again he shows how it exists independently of, and in addition to, all the other powers that he had been discussing. In addition, there is repulsion as well as attraction:

Now there are some curious bodies in nature (of which I

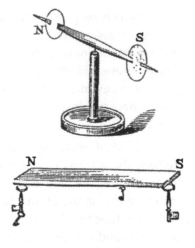

Magnets.

have two specimens on the table) which are called *magnets* or *loadstones*—ores of iron, of which there is a great deal sent from Sweden. They have the attraction of gravitation, and attraction of cohesion, and certain chemical attraction, but they also have a great attractive power, for this little key is held up by this stone. Now, that is not chemical attraction; it is not the attraction of chemical affinity, or of aggregation of particles, or of cohesion, or of electricity for it will not attract this ball if I bring it near it, but it is a separate and dual attraction—and, what is more, one which is not readily removed from the substance, for it has existed in it for ages and ages in the bowels of the earth.

Here is a little magnet, and I have coloured the ends differently, so that you may distinguish one from the other. Now this end (S) of the magnet attracts the *uncoloured* end of the little magnet. You see it pulls it towards it with great power, and as I carry it round, the uncoloured end still follows. But now, if I gradually bring the middle of the bar magnet opposite the uncoloured end of the needle, it has no effect upon it, either of attraction or repulsion, until, as I come to the opposite extremity (N), you see that it is the *coloured* end of the needle which is pulled towards it. We are not therefore, dealing with two kinds of power, attracting different ends of the magnet—a double power, already existing in these bodies, which takes up the form of attraction and repulsion. And now, when I put up this label with the word MAGNETISM, you will understand that it is to express this double power.

As the lectures draw to their close Faraday turns to the deepest question of all, whether and in what way, the various forces whose effects he has been describing are related to one another, and whether all may be different manifestations of some other primal force. It is quite salutary to remember that these matters were being raised in simple direct language before an audience whose average age was probably about 14:

We have frequently seen, during the course of these lectures, that one of those powers or forces of matter, of which I have written the names on that board, has produced results which are due to the action of some other force. Thus, you have seen the force of electricity acting in other ways than in attracting. You have also seen it combine matters together, or disunite them, by means of its action on the chemical force, and in this case, therefore, you have an instance in which these two powers are related. But we have other and deeper relations that these. We have not merely to see how it is that one power affects another—how the force of heat affects chemical affinity, and so forth—but we must try and comprehend what relation they bear to each other, and how these powers may be changed one into the others. And it will today require all my care, and your care too, to make this clear to your minds. I shall be obliged to confine myself to one or two instances, because, to take in the whole extent of this mutual relation and conversion of forces, would surpass the human intellect.

Faraday's own driving conviction is evident: the forces he is talking about *are* all connected. That modesty that is such a charming and delightful feature of his character prevents him from pointing out that more that one of the connections in question was established by his own efforts. But hope and expectation continue to guide the search:

> I have here given you all the illustrations that time will permit me to shew you of chemical affinity producing electricity, and electricity again becoming chemical affinity. Let that suffice for the present, and let us now go a little deeper into the subject of this chemical force, or this electricity. Which shall I name first—the one producing the other in a variety of ways? These forces are also wonderful in their power of producing another of the forces we have been considering, namely, that of magnetism, and you know that it is only of late years, and long since I was born, that the discovery of the relations of these two forces of electricity and chemical affinity to produce magnetism have become known.
>
> Philosophers had been suspecting this affinity for a long time and had long had great hopes of success for, in the pursuit of science, we first start with hopes and expectations. These we realise and establish, never again to be lost, and upon them we found new expectations of further discoveries, and so go on pursuing, realising, establishing, and founding new hopes again and again.

Generating a magnetic field by passing an electric current through a wire gives instant evidence of the connection between magnetism and electricity, and the point is rendered even more evident by repeating the experiment performed in an earlier lecture, making a solid bridge out of iron filings attracted to a magnet. Only this time the magnet is not a magnetised lump of iron, but a coil of copper, an element that is not magnetic in itself, but can become the source of a magnetic field when it carries a current. Furthermore, the current has been produced by a battery, so the electricity arises from chemical action:

> I have here a quantity of wire, which has been wound into a spiral, and this will affect the magnetic needle in a very

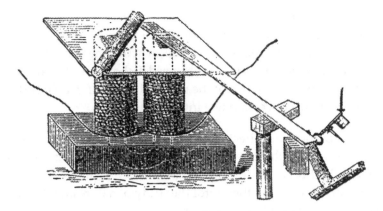

An electromagnet.

curious manner because, owing to its shape, it will act very like a real magnet. The copper spiral has no power over that magnetic needle at present, but if I cause the electric current to circulate through it, by bringing the two ends of the battery in contact with the ends of the wire which forms the spiral, what will happen? Why, one end of the needle is most powerfully drawn to it, and if I take the other end of the needle, it is repelled. So you see I have produced exactly the same phenomena as I had with the bar magnet, one end attracting, and the other repelling. Is not this then curious to see that we can construct a magnet of copper?

I will repeat the experiment which you saw the other day, of building up a bridge of iron nails. The contact is now made, and the current is going through; it is now a powerful magnet. Here are the iron nails which we had the other day, and now I have brought this magnet near them, they are clinging so hard that I can scarcely move them with my hand. But when the contact is broken, see how they fall. What can show you better than such an experiment as this the magnetic attraction with which we have endowed these portions of iron?

What, then, can surpass these evidences of the change

of chemical force into electricity, and electricity into magnetism? I might show you many other experiments whereby
I could obtain electricity and chemical action, heat and light,
from a magnet, but what more need I show you to prove
the universal correlation of the physical forces of matter, and
their mutual conversion one into another?

The peroration, directed at the grown ups in the audience,
refers to the fact that the lectures had to be postponed because
Faraday had been ill. But his final remarks of all could be for
young and old alike as they extol the study of nature and its laws,
among the highest of human callings:

And now, let us give place as juveniles, to the respect we
owe to our elders; and for a time let me address myself
to those of our seniors who have honoured me with their
presence during these lectures. I wish to claim this moment
for the purpose of tendering our thanks to them, and my
thanks to you all, for the way in which you have borne the
inconvenience that I at first subjected you to. I hope that the
insight which you have here gained into some of the laws by
which the universe is governed may be the occasion of some
amongst you turning your attention to these subjects. For
what study is there more fitted to the mind of man than that of
physical sciences? And what is there more capable of giving
him an insight into the actions of those laws, a knowledge
of which gives interest to the most trifling phenomenon of
nature, and make the observing student find—

'—tongues in trees, books in the running brooks,
Sermons in stones, and good in everything?'

Chapter 11

Honour and Recognition

Faraday's views about the rights of a researcher to receive due recognition for the novelty of an idea or an experimental result were founded quite firmly on his own personal moral principles. He also observed the behaviour of his fellow scientists (and even the world at large) with a certain degree of disapprobation, touched (dare one say?) from time to time with a mite of sanctimoniousness. Addressing the French physicist André-Marie Ampère, whose fame has lasted well into the present day (not least through the name of the unit of electric current), Faraday alludes to an unfortunate episode in his own relations with his mentor Humphry Davy:

> I am sorry to find that you experience an unworthy opposition to the fair and high claim you have to the approbation and thanks of your fellow philosophers. This, however, you can hardly wonder at. I do not know what it is, or by whom exerted in your case, but I never yet, even in my short time, knew a man to do anything eminent, or become worthy of distinction, without becoming at the same time obnoxious to the cavils and rude encounters of envious men. Little as I have done, I have experienced it and that, too, where I least expected it.

In similar vein, priority was an issue that greatly preoccupied Faraday. Determining to pursue science (or natural philosophy) from an early age because of what he conceived as its high and noble objectives, he remained unreconciled to the evidence he frequently encountered of venality and misconduct which prevails in this as in any other field of human activity. Thus to James David Forbes, Professor of Natural Philosophy at Edinburgh University he wrote in 1832:

> You speak of the transmission of information and the lesson it affords to discoverers. But is it not very annoying that one may not talk of a matter to one's most intimate friend lest it should be misinterpreted, or perhaps given to another or, as has happened to myself (in other cases, *not* the present), be actually stolen? But I must not allow my recollections to dwell on these things, although they sometimes almost induce me to give up the pursuit of science for, exalted and noble as it is in itself, in its outward appearance it frequently presents, to the private knowledge of him who pursues it, quite as much that is degraded and base.

The custom in the nineteenth century, as indeed it remains today, is for candidates aspiring to a post to ask eminent people of their acquaintance to write letters of recommendation (what he called certificates) on their behalf, drawing attention to their virtues and achievements. Faraday was loathe to write such letters, perhaps in part because he was aware how valued his recommendation would have been, as in a reply to Francis Booth, the Secretary of the Linnean Society, in 1837:

> That I cannot give a testimonial to my friend Phillips is not merely a matter of general reluctance to certify, but of principle for, refusing all, I am obliged to refuse each, and I refuse all because, having no confidence in certificates, I wish to have no association in any way with them. In the present particular case too, if I certify for one I should have to certify for two others also. But when I am asked by an authority concerned in forming the decision what I think of a candidate, I cannot refuse to answer.

One form of acknowledgement of his achievements that Faraday himself received in abundance was honorary membership of academies and learned societies. As a result (and also because of the principles governing his own view about attribution of credit for such achievements) he had decided opinions about the conditions under which it was right to accept such honours. In brief, they acknowledged distinction but (in Faraday's mind) carried no obligation towards the institution granting them. In 1839, Abraham Follett Osler, the meteorologist, received the following from Faraday:

> I will tell you what my opinions of Honorary Memberships are. I look upon them as marks of approbation bestowed by bodies upon such individuals as they may conceive are worthy of such reward, and I think the individuals can hardly be so vain as to balance whether they will accept such praise or not: it is for them thankfully to receive. Under such circumstances you will judge what my feelings would be in the case which you do me favour to put.
>
> On the other hand I am obliged to tell you that I have nothing I can return for such honours, for all my time and attention is devoted in channels already determined, to the pursuit of science. Nor can I leave them in any degree for a new cause, or another body than those I am already engaged with.

Five years later his position on the matter remained unchanged, and the advice offered to Thomas Andrews, Professor of Chemistry at the Belfast Academic Institute, was the same:

> As to the particular point of your letter, about which you honour me by asking my advice, I have no advice to give but I have a strong feeling on the matter, and will tell you what I should do. I have always felt that there is something degrading in offering rewards for intellectual exertion, and that Societies or Academies (or even Kings & Emperors) should mingle in the matter does not remove the degradation. For the feeling which is hurt is a point above

their condition and belongs to the respect which a man owes to himself. With this feeling, I have never since I was a boy aimed at any such prize (or even if, as in your case, they came near me) have allowed them to move me from my course, and I have always contended that such results will never move the men who are most worthy of reward. Still, I think rewards and Honours *good* if properly distributed, but they should be given for what a man has done and not offered for what he is to do, or else talent must be considered as a thing marketable and to be bought and sold. And then down falls that high tone of mind which is the best excitement to a man of power, and will make him do more than any common place reward, and they give it, not as a moving impulse to him, but to all those who by the reward are led to look to that man for an example. If I were you, therefore, I should go on my way and discover and publish (if I can) but, having done that, I see no objection (as the time draws nigh) to send copies of the papers to the Academy, or even such an account of them as may be considered fit and in doing so I should think I was paying a fit mark of respect to the Academy, and giving them the opportunity of marking their sense of what had been done, if they saw fit. But I would not depart from my own high position (I mean as respects feeling), for any reward they could give. Excuse my freedom. I have no time to dot ii or cross tt or punctuate. I hope you will find out the meaning.

And, as a final remark on the question of honours (this time of a public kind), the brief reply to the owner of a chemist's shop in Brighton who had addressed him as 'Sir' is plain in the extreme:

You do me too much honour in your address, for I am plain Mr Faraday and shall so remain.

It was Faraday's settled opinion that governments should, for their own sake, honour those who, by pursuing scientific enquiry, illuminated the underlying principles that govern the natural world. Still, knowing that governments normally give awards to their political allies and supporters, he was wary of honours such as knighthoods that did not distinguish in their title the reason for which the honour was granted. Practical as ever, Faraday's advice in 1854 to Lord Wrottesley, the Secretary of the Parliamentary Committee of the British Association, who had solicited his opinion, was to reward talent, not through appellations, but by employing such individuals in the service of the country. For Faraday himself, his life at the Royal Institution had been reward enough:

> I feel unfit to give a deliberate opinion on the course it might be advisable for the government to pursue, if it were anxious to improve the position of science and its cultivators in our country. My course of life, and the circumstances which make it a happy one for me, are not those of persons who conform to the usages and habits of society. Through the kindness of all, from my sovereign downward, I have that which supplies all my need and in respect of honours, I have, as a scientific man, received from foreign countries and sovereigns those which, belonging to very limited and select classes, surpass in my opinion any thing that is in the power of my own to bestow.

> I cannot say that I have not valued such distinctions. On the contrary, I esteem them very highly, but I do not think I have ever worked for, or sought after them. Even were such to be now created here, the time is past when these would possess any attraction for me and you will see, therefore, how unfit I am, upon the strength of any personal motive or feeling, to judge of what might be influential upon the minds of others. Nevertheless, I will make one or two remarks which have often occurred to my mind.

> Without thinking of the effect it might have upon distinguished men of science, or upon the minds of those who, stimulated to exertion, might become distinguished, I do think that a government should *for its own sake* honour the

men who do honour and service to the country. I refer now to honours only, not to beneficial rewards; of such honours I think there are none. Knighthoods and baronetcies are sometimes conferred with such intentions, but I think them utterly unfit for that purpose. Instead of conferring distinction, they confound the man who is one of twenty, or perhaps fifty, with hundreds of others. They depress rather than exalt him, for they tend to lower the especial distinctive mind to the commonplaces of society. An intelligent country ought to recognize the scientific men among its peoples as a class. If honours are conferred upon eminence in any class, as that of the law or the army, they should be in this also. The aristocracy of the class should have other distinctions than those of lowly and highborn, rich and poor, yet they should be such as to be worthy of those whom the sovereign and the country should delight to honour and, being rendered very desirable and even enviable in the eyes of the aristocracy by birth, should be unattainable except to those of science. Thus much I think the government and the country ought to do, for their own sake and the good of science, more than for the sake of men who might be thought worthy of such distinction. The latter have attained to fit their place, whether the community at large recognize it or not.

But besides that, and as a matter of reward and encouragement to those who have not yet risen to great distinction, I think the government should, in the very many cases which come before it having a relation to scientific knowledge, employ men who pursue science, provided they are also men of business. This is perhaps now done to some extent, but to nothing like the degree which is practicable with advantage to all parties. The right means can not have occurred to a government which has not yet learned to approach and distinguish the class as a whole.

Chapter 12

Public Affairs: Consultant and Advocate

In the early years of the nineteenth century the Royal Institution was a natural focus of what would nowadays be called scientific consultancy. Problems were brought to the notice of the Royal Institution's small staff, and work was conducted to solve them. The fees paid for such work constituted an important part of the organisation's income. As a junior member of the laboratory many of these tasks fell to Faraday, but with increasing eminence came invitations to take a leading part in official enquiries, such as that into the causes of the disastrous explosion and fire at Haswell Colliery in 1844. In a final phase, as a revered figure he expressed his own views spontaneously on matters of public interest as diverse as the place of science in education and the urgent need to clean up the River Thames.

Throughout his career, and increasingly as his scientific stature became greater, Faraday was called on to give advice about technological problems that arose from the rapid progress of industrialisation across the country. As it became more widely recognised that these problems, such as the fires that sometimes broke out spontaneously in the coal stocks on steam ships, could be explained (and hence tackled) by recourse to basic knowledge of chemistry and physics, so the Royal Institution, as a major repository for such knowledge, was a natural place to look for solutions. In the example of storing coal on ships, the Holyhead Road Committee gained some very mundane practical advice, backed by Faraday's knowledge of the chemistry of iron pyrites. The authoritative tone of the memorandum is worth noting: Faraday was 31:

> The proper selection and state of the coals used on board steam vessels, especially when one very great object is to secure a regular and short passage, whatever the state of the weather may be, is very important. I am not, however, sufficiently acquainted with the varieties of coal raised in this country to state the particular qualities which distinguish them, or to point out the best for this particular purpose. As, however, it is not practicable to have access to any coal at pleasure, all that can be done is to ascertain the qualities of those within reach, and choose the best.

> All coal containing pyrites should be rejected, or at least the purest kind selected. Mention has been made before the Committee of the spontaneous inflammation of coal on board a steam vessel. This is an effect which, I believe, not unfrequently takes place at the pit mouth, and when produced, it is due to the pyrites or sulphuret of iron the coal contains. When large masses of coal containing this substance are exposed to air and moisture, those agents sometimes act on the pyrites, causing the oxidation of the iron in them, and liberating sulphur and combustible aërial products. And the heat produced by this action in the mass sometimes rises so high as to fire the sulphur and sulphuretted gases, and these inflame the coal, so that the combustion of the mass is produced without the approximation of any ignited body.

> The coals in steam vessels are generally, I believe, put into boxes at the sides of the place where the engine stands, and I do not think that with moderate caution they can fire there, *i.e.* provided the communication between them and the furnaces be prevented; for their masses are comparatively

so small, and they remain for so very short a time together undisturbed, that it is not probable the mere action of air and moisture on the pyrites should inflame them. It is true, the elevation of temperature in the place in which they are stowed, caused by the vicinity of the furnaces, will be favourable to such an effect, but I do not think its influence would be dangerous, except with very bad coals indeed. It would, however, be easy at all times to know the state of the coals, by having one, two or three iron pipes going across the boxes in which they are put, and tendency in the coals to heat would heat pipes also, and might be detected by a thermometer being placed in them.

Nevertheless, there are signs that at this stage of his life, Faraday was finding his analytical and consultancy work irksome. To the French chemist André-Marie Ampère, who he had met in Paris with Davy in 1813 he wrote in 1823:

Considering the very subordinate situation I hold here and the little encouragement which circumstances hold out to me, I have more than once been tempted to resign scientific pursuits altogether, but then the remembrance of such letters and expressions as yours cheers me again and I struggle on, in hopes of getting results at one time or another that shall (by their novelty or interest) raise me into a more liberal and active sphere.

As a person whose existence was focused towards getting on with the job, and acutely (even obsessively) conscious that time (his 'only estate') was to be used effectively, Faraday was naturally impatient with time wasters, especially people who had been nominated to committees for reasons other than that of expertise or commitment. For example, following a meeting of the Royal Society Council in 1833, which he clearly found frustrating, he shared his view with John William Lubbock, who had been in the Chair:

With respect to committees, as you would perceive I am very jealous of their formation. I mean *working committees*. I think business is always better done by few than by many. I think also the working few ought not to be embarrassed by the idle many and, further, I think the idle many ought not to be honoured by association with the working few. I do not think that my patience has ever come nearer to an end than when compelled to hear (in the examination of witnesses etc, etc in committee) the long rambling malapropos enquiries of members who still have nothing in consequence to propose that shall advance the business. But in all this too, I will promise to behave as well as I can.

In public affairs he is seen at his incisive best when making written reports on precise issues, as for example the following advice to James Cosmo Melvill, Chief Secretary of the East India Company, on the efficiency of lightning conductors:

Sir,

I have the honor to acknowledge your letter and the papers and, having read the latter, beg leave to state that my opinion is in favour of lightning conductors. It is no doubt true that low rounded buildings, such as I understand the powder magazines in the East Indies to be, are but very little liable to be struck by Lightning, but then, if they are struck, the destruction and injury may be very great. It is also, I think, very probable that a lightning conductor may under *certain circumstances* cause an electric discharge to take place where none would have occurred (no conductor being present) though on the other hand, there is some evidence to show that conductors cause a diminution in the number of electric discharges to the earth at a given place. It is also very certain that a badly erected conductor is worse than none, and may cause great injury. But notwithstanding these considerations I have the strongest conviction in my mind that conductors *well applied* are perfect defenders of buildings from harm by lightning.

I would certainly recommend copper conductors instead of iron, for the former metal conducts electricity almost seven times better than the latter. When struck, it not only conducts the shock much better, but in the predetermination of the stroke, it determines more of the electricity to itself than otherwise would fall upon it and therefore tends, in any case of a divided shock, to leave less to fall elsewhere in its neighbourhood.

I should prefer them pointed. I should not put them far from the building at their upper extremity, or in their courses downwards, but the part that is underground I should turn from the building in its course through the earth, and take especial care by plates of copper to make its contact with the moist earth extensive and good.

Conductors should be of certain height in relation to the roof or summit of the building to be defended. A lightning rod rising ten feet above any part of the roof or chimneys of a house might defend that house perfectly if close to it, but not if ten feet from it. A rod rising fifteen feet above the highest parts of the roof would be more sure than one of ten feet. A rod projecting ten feet which would protect a building of a certain horizontal extent might not protect a building ten feet wider, and a lightning rod has been considered as able to protect objects perfectly when they are not more than twice the distance from it of *height above them*. But for this to hold true, these objects should not be themselves parts of large masses of metal approaching by their position and connection to the character of bad lightning conductors.

I have no fear of lateral discharge from a *well managed* conductor. As far as I understand lateral discharge, it is always a discharge from the conductor itself. It might be very serious from a badly managed conductor (and in fact makes them worse than nothing) but with a good lightning rod it can be but small, and then not to badly conducting matter as wood or stone, but only to neighbouring masses of good-conducting matter as the metals which either ought not to be there (or if they are necessarily present) ought to be in metallic communication with the lightning conductor itself. I am not aware that lateral discharge can take place *within* a building when a lightning conductor outside is struck, except there be portions of metal (as bolt wires or bolts etc) which may form an interrupted conducting train from the conductor to the interior. It is true that cases which came under the denomination of returning stroke might perhaps produce a spark in the interior of a building, but the phenomena of a returning stroke cannot occur at the place where the lightning strikes a conductor.

In my opinion a *good* conductor, well connected with the earth, cannot do harm to a building under its protection, i.e. though it may induce a discharge upon itself, it cannot induce a discharge on the building: and the discharge on itself cannot give rise to any secondary effects which are likely to place the building in more danger than it would have been subject to had the conductor not been there.

As he gave lectures to the cadets at the Royal Military Academy at Woolwich, Faraday was required to conduct examinations. But from time to time the cadets prompted one another on the answers to his questions, behaviour which (in a letter to William Gravatt, the Inspector of the Academy) drew forth all his forces of moral indignation. Still, in this situation as in so many others, he had a practical solution to hand.

I write to ask your advice.   The matter regards the examination which, as it is intended to show who has taken advantage of the instruction in the lectures and who has not, should not be thwarted or become a mere matter of form. I think you will be aware that I have several times noticed and disapproved of the practice of some to prompt others in their answers and requested that it should not be continued. I certainly consider it a degree of mental degradation in a young man, when he is willing to do that in secret which he does not do openly, and knows he ought not to do at all, and such is the case of the prompter. As to the prompted, I

Faraday holding a bar magnet, photograph published in 1859 in the series *Photographic Portraits of Living Celebrities.*

have more respect by far for the person who would openly acknowledge his inability to answer a question, than for him who would use the secretly imparted knowledge of another, and so claim that credit *which is not his due.* I intended to consult you as to what you would think of the plan of calling these gentlemen whom I intended to examine from their seats after lecture, and making them stand separately from the seats, and from each other, during the examination. I should be very sorry indeed to do so, as I should consider such a proceeding a great degradation. I sincerely hope that no such plan will be necessary: but on the other hand that the practice complained of must be stopped, whatever the means may be necessary for that purpose, is a point on which I am sure you will agree with me.

Two letters to the *Times* newspaper encapsulate Faraday's public reaction to issues that had become matters of public concern towards the end of his life. In the early 1850s a craze for the paranormal grew up in London, which manifested itself in sessions of 'table turning', a phenomenon that recurred periodically in the early years of this century. The notion that nature's laws could be suspended by occult means was deeply offensive to Faraday's world view, but his public denunciation of the deceptions involved is based firmly on his skills as an observer and experimenter. The following appeared in the *Times* on 30 June 1853:

> SIR,—I have recently been engaged in the investigation of table-turning. I should be sorry that you should suppose I thought this necessary on my own account, for my conclusion respecting its nature was soon arrived at, and is not changed. But I have been so often misquoted, and applications to me for an opinion are so numerous that I hoped, if I enabled myself by experiment to give a strong one, you would consent to convey it to all persons interested in the matter. The effect produced by table-turners has been referred to electricity, to magnetism, to attraction, to some unknown or hitherto unrecognized physical power able to effect inanimate bodies—to the revolution of the earth, and even to diabolical or supernatural agency. The natural philosopher can investigate all these supposed causes but the last. That must, to him, be too much connected with credulity or superstition to require any attention on his part. The investigation would be too long in description to obtain a place in your columns. I therefore propose asking admission for that into the 'Athenæum' of next Saturday, and propose here to give the general result.
>
> Believing that the first cause assigned—namely, a *quasi*

involuntary muscular action (for the effect is with many subjects to the wish or will)—was the true cause, the first point was to prevent the mind of the turner having an undue influence over the effects produced in relation to the nature of the substances employed. A bundle of plates, consisting of sand-paper, millboard, glue, glass, plastic clay, tinfoil, cardboard, gutta-percha, vulcanized caoutchouc, wood, and resinous cement, was therefore made up and tied together and, being placed on a table, under the hand of a turner, did not prevent the transmission of the power. The table turned or moved exactly as if the bundle had been away, to the full satisfaction of all present. The experiment was repeated, with various substances and persons, and at various times, with constant *success*, and henceforth no objection could be taken as to the use of these substances in the construction of apparatus.

The next point was to determine the place and source of motion, *i.e.* whether the table moved the hand, or the hand moved the table, and for this purpose indicators were constructed. One of these consisted of a light lever, having its fulcrum on the table, its short arm attached to a pin fixed on a card board which could slip on the surface of the table, and its long arm projecting as an index of motion. It is evident that if the experimenter willed the table to move towards the left, and it did so move before the hands, placed at the time on the cardboard, then the index would move to the left also, the fulcrum going with the table. If the hands involuntarily moved towards the left without the table, the index would go towards the right and, if neither table nor hands moved, the index would itself remain immoveable. The result was, that when the parties saw the index it remained very steady. When it was hidden from them, or they looked away from it, it wavered about, though they believed that they always pressed directly downwards and, when the table did not move, there was still a resultant of hand force in the direction in which it was wished the table should move, which, however, was exercised quite unwittingly by the party operating. The resultant it is which, in the course of the waiting time, while the fingers and hands become stiff, numb, and insensible by continued pressure, grows up to an amount sufficient to move the table *or* the substances pressed upon. But the most valuable effect of this test-apparatus (which was afterwards made more perfect and independent of the table) is the corrective power it possesses over the mind of the table-turner.

As soon as the index is placed before the most earnest, and they perceive—as in my presence they have always done— that it tells truly whether they are pressing downwards only or obliquely, then all effects of table-turning cease, even though the parties persevere, earnestly desiring motion, till they become weary and worn out. No prompting or checking of the hands is needed—the power is gone; and this only because the parties are made conscious of what they are really doing mechanically, and so are unable unwittingly to deceive themselves. I know that some may say that it is the cardboard next the fingers which moves first, and that it both drags the table, and also the table-turner with it. All I have to reply is that the cardboard may in practice be reduced to a thin sheet of paper weighing only a few grains, or to a piece of goldbeaters' skin, or even the end of the lever, and (in principle) to the very cuticle of the finger itself. Then the results that follow are too absurd to be admitted: the table becomes an incumbrance, and a person holding out the fingers in the air, either naked or tipped with goldbeaters' skin or cardboard, ought to be drawn around the room etc. But I refrain from considering imaginary yet consequent results which have nothing philosophical or real in them. I have been happy thus far in meeting with the most honourable and candid (though most sanguine) persons, and I believe the mental check which I propose will be available in the hands of all who desire truly to investigate the philosophy of the subject and, being content to resign expectation, wish only to be led by the facts and the truth of nature. As I am unable, even at present, to answer all the letters that come to me regarding this matter, perhaps you will allow me to prevent any increase by saying that my apparatus may be seen at the shop of the philosophical instrument maker Newman, 122 Regent Street.

Permit me to say, before concluding, that I have been greatly startled by the revelation which this purely physical subject has made of the condition of the public mind. No doubt there are many persons who have formed a right judgement or used a cautious reserve, for I know several such, and public communications have shown it to be so. But their number is almost as nothing to the great body who have believed and borne testimony, as I think, in the cause of error. I do not here refer to the distinction of those who agree with me and those who differ. By the 'great body', I mean such as reject all consideration of the equality of cause and effect, who refer the results to electricity and magnetism, yet know nothing of the laws of these forces—or to attraction, yet show no phenomena of pure attractive power—or to the rotation of the earth, as if the earth revolved round the leg of a table— or to some unrecognized physical force, without inquiring whether the known forces are not sufficient—or who even refer them to diabolical or supernatural agency, rather than suspend their judgement, or acknowledge to themselves that they are not learned enough in these matters to decide on the nature of the action. I think the system of education that could leave the mental condition of the public body in the state in which this subject found it, must have been greatly deficient in some very important principle.

In contrast, the matter of cleaning the River Thames of the putrescent effluent deposited in it as it flowed through London was a legitimate object for public action. Once again, Faraday's expedient to see how opaque the water was shows us how sure was his grasp of simple experiment. One is reminded of a more recent example: Richard Feynman at the enquiry into the Challenger disaster. On 7 July 1855 the *Times* carried the following:

SIR,

I traversed this day by steamboat the space between London and Hungerford Bridges, between half-past one and two o'clock. It was low water, and I think that tide must have been near the turn. The appearance and smell of the water forced themselves at once on my attention. The whole of the river was an opaque brown fluid. In order to assess the degree of opacity, I tore up some white cards into pieces, and then moistened them, so as to make them sink easily below the surface, and then dropped some of these pieces into the water at every pier the boat came to. Before they had sunk an inch below the surface they were undistinguishable, though the sun shone brightly at the time, and when the pieces fell edgeways the lower part was hidden from sight before the upper part was under water.

This happened at St. Paul's Wharf, Blackfriars Bridge, Temple Wharf, Southwark Bridge and Hungerford, and I have no doubt would have occurred further up and down the river. Near the bridges the feculence rolled up in clouds so dense that they were visible at the surface even in water of this kind.

The smell was very bad, and common to the whole of the water. It was the same as that which now comes up from the gully holes in the streets. The whole river was for the time a real sewer. Having just returned from the country air, I was perhaps more affected by it than others, but I do not think that I could have gone on to Lambeth or Chelsea, and I was glad to enter the streets for an atmosphere which, except near the sink-holes, I found much sweeter than on the river.

I have thought it a duty to record these facts, that they may be brought to the attention of those who exercise power, or have responsibility in relation to the condition of our river. There is nothing figurative in the words I have employed, or any approach to exaggeration. They are the simple truth.

Cartoon from *Punch* (21 July 1855) showing Faraday on the Thames.

If there be sufficient authority to remove a putrescent pond from the neighbourhood of a few simple dwellings, surely the river which flows for so many miles through London ought not to be allowed to become a fermenting sewer. The condition in which I saw the Thames may perhaps be considered as exceptional, but it ought to be an impossible state, instead of which, I fear it is rapidly becoming the general condition. If we neglect this subject, we cannot expect to do so with impunity. Nor ought we to be surprised if, ere many years are over, a season give us sad proof of the folly of our carelessness.

I am, Sir, your obedient servant,

M. FARADAY

During the 1850s Faraday's standing in the scientific community and as a public figure, not to mention the strongly held convictions that his career had established in his mind, led him to be drawn increasingly into the debate about the proper place that science should occupy in the education and training of young people. Two lectures, in particular, encapsulate his views. The first takes its starting point from the perennial and continuing debate about the role of science as a basis for technological progress, but widens the argument to consider what the study of science can bring to education. As ever, Faraday's point of departure was a concrete example, as can be seen from the text of the lecture, presented as a Friday Evening Discourse at the Royal Institution in 1854: it was called 'Wheatstone's Electric Telegraph's Relation to Science (being an argument in favour of the full recognition of Science as a branch of Education)':

> If the term education may be understood in so large a sense as to include all that belongs to the improvement of the mind, either by the acquisition of the knowledge of others or by increase of it through its own exertions, we learn by them what is the kind of education science offers to man. It teaches us to be *neglectful* of nothing; not to despise the small beginnings, for they precede of necessity all great things in the knowledge of science, either pure or applied. It teaches a continual comparison of the *small* and *great*, and that under differences almost approaching the infinite: for the small as often contains the great in principle as the great does the small, and thus the mind becomes comprehensive. It teaches us to deduce principles carefully, to hold them firmly, or to suspend the judgement; to discover and obey *law*, and by it to be bold in applying to the greatest what we know of the smallest. It teaches us first by tutors and books to learn that

which is already known to others, and then by the light and
methods which belong to science to learn for ourselves and
for others, so making a fruitful return to man in the future for
that which we have obtained from the men of the past. Bacon,
in his instruction, tells us that the scientific student ought
not to be as the ant, who gathers merely, nor as the spider
who spins from her own bowels, but rather as the bee who
both gathers and produces. All this is true of the teaching
afforded by any part of the physical science. Electricity is
often called wonderful, beautiful, but it is so only in common
with the other forces of nature. The beauty of electricity, or
of any other force, is not that the power is mysterious and
unexpected, touching every sense at unawares in turn, but
that it is under *law*, and that the taught intellect can even
now govern it largely. The human mind is placed above, not
beneath it, and it is in such a point of view that the mental
education afforded by science is rendered supereminent in
dignity, in practical application, and utility, for, by enabling
the mind to apply the natural power through law, it conveys
the gifts of God to man.

The apotheosis of Faraday's public utterances on the place
of sciences in a general education took place on an auspicious
occasion in May 1854 when he spoke at the Royal Institution to
an audience that included Prince Albert, the Prince Consort. The
lecture was one of a series by eminent savants, which stemmed
from Faraday's complaint, in his letter about table turning, about
the lack of scientific education for the public at large. The title of
the oration was short: 'On Mental Education', but before entering
into the mainstream of his argument Faraday gave one of the
clearest distinctions found in any of his published work between
knowledge arrived at through the pursuit of scientific enquiry and
beliefs that resulted from religious revelation, and were adhered
to by faith. (It is worth noting in parenthesis that the reference to
the Godhead in the following extract is a direct quotation from St
Paul's Epistle to the Romans):

Before entering upon the subject, I must take one distinction which, however it may appear to others, is to me of the utmost importance. High as man is placed above the creatures around him, there is a higher and far more exalted position within his view, and the ways are infinite in which he occupies his thoughts about the fears, or hopes, or expectations of a future life. I believe that the truth of that future cannot be brought to his knowledge by any exertion of his mental powers, however exalted they may be; that it is made known to him by other teaching than his own, and is received through simple belief of the testimony given. Let no one suppose for a moment that the self-education I am about to commend in respect of the things of this life, extends to any considerations of the hope set before us, as if man by reasoning could find out God. It would be improper here to enter upon this subject further than to claim an absolute distinction between religious and ordinary belief. I shall be reproached with the weakness of refusing to apply those mental operations which I think good in respect of high things to the very highest. I am content to bear the reproach. Yet, even in earthly matters, I believe that the invisible things of HIM from the creation of the world are clearly seen, being understood by the things that are made, even His eternal power and Godhead; and I have never seen anything incompatible between those things of man which can be known by the spirit of man which is within him, and those higher things concerning his future, which he cannot know by that spirit.

Having clearly set the boundaries beyond which he would not countenance extending the mode of enquiry to which he was so deeply attached as a means of obtaining reliable information about the natural world, Faraday proceeds, in this lecture, to demonstrate how unreliable intuition can be. Mental education is construed as all that knowledge obtained in our earliest years, storing it up to become the unconscious source of expectation that we call intuition. Faraday's case is that such intuition can often lead to false conclusions, and he illustrates his point with a very simple demonstration that anyone can repeat and verify for themselves:

As I believe that a very large proportion of the errors we make in judgement is a simple and direct result of our perfectly unconscious state, and think that a demonstration of the liabilities we are subject to would aid greatly in providing a remedy, I will proceed first to a few illustrations of a physical nature. Nothing can better supply them than the limitations we derive from our senses. To them we trust directly, by them we become acquainted with external things and gain the power of increasing and varying facts, upon which we entirely depend. Our sense-perceptions are wonderful. Even in the observant but unreflective infant, they soon produce a result which looks like intuition, because of its perfection. Coming to the mind as so many data, they are stored up, and without our being conscious, are ever after used in like circumstances in forming our judgement. And is it not wonderful that man should be accustomed to trust them without examination? Nevertheless, the result is the effect of education. The mind has to be instructed with regard to the senses and their intimations through every step of life, and where the instruction is imperfect, it is astonishing how soon and how much their evidence fails us. Yet, in the latter years of life, we do not consider this matter but, having obtained the ordinary teaching sufficient for ordinary purposes, we venture to judge of things which are extraordinary for the first time, and almost always with the more assurance as our powers of observation are less educated.

Consider the following case of a physical impression derived from the sense of touch, which can be examined and verified at pleasure: if the hands can be brought towards each other so that the tips of the corresponding fingers touch, the end of any finger may be considered as an object to be felt by the opposed finger; thus the two middle fingers may for the present be so viewed. If the attention be directed to them, no difficulty will be experienced in moving each lightly in a circle round the tip of the other, so that they shall each feel the opposite, and the motion may be either in one direction or the other—looking at the fingers, or with eyes employed elsewhere—or with the remaining fingers touching quiescently, or moving in a like direction. All is easy, because each finger is employed in the ordinary or educated manner whilst obeying the will, and whilst communicating through the sentient organ with the brain. But turn the hands halfway round, so that their backs shall be towards each other, and then, crossing them at the wrists, again bring the like fingers into contact at the tips. If it be now desired to move the extremities of the middle fingers round each other, or to follow the contour of one finger by the tip of the opposed one, all sorts of confusion in the motion will ensue, and as the finger of one hand tries, under the instruction of the will, to move in one course, the touched finger will convey an intimation that it is moving in another. If all the fingers move at once, all will be in confusion, the ease and simplicity of the first case having entirely disappeared. If, after some considerable trial, familiarity with the new circumstances have removed part of the uncertainty, then, crossing the hands at the opposite sides of the wrists will renew it. These contrary results are dependent not on any change in the nature of the sentient indication, or of the surfaces or substances which the sense has to deal with, but upon the trifling circumstance of a little variation from the direction in which the sentient organs of these parts are usually exerted. And they show to what an extraordinary extent our interpretations of the sense impressions depend upon the experience, *i.e.* the education which they have previously received, and their great inability to aid us at once in circumstances which are entirely new.

If our intuitions lead us into false expectations about how the world behaves, it is science that comes to our rescue, being the distillation of knowledge arrived at by carefully designed experiments, and subjected to vigorous criticism by successive generations. The ultimate distillation of such knowledge takes the form of what Faraday calls 'the laws of nature'—the most reliable summary that we have, yet eternally provisional:

> The *laws of nature*, as we understand them, are the foundation of our knowledge in natural things. So much as we know of them has been developed by the successive energies of the highest intellects, exerted through many ages. After a most rigid and scrutinizing examination upon principle and trial, a definite expression has been given to them. They have become, as it were, our belief or trust. From day to day we still examine and test our expressions of them. We have no interest in their retention if erroneous. On the contrary, the greatest discovery a man could make would be to prove that one of these accepted laws was erroneous, and his greatest honour would be the discovery. Neither should there be any desire to retain the former expression, for we know that the new or the amended law would be far more productive in results, would greatly increase our intellectual acquisitions, and would prove an abundant source of fresh delight to the mind.
>
> When we think of the laws of nature (which by continued observation have become known to us) as the proper tests to which any new fact or our theoretical representation of it should in the first place be subjected, let us contemplate their assured and large character. Let us go out into the field and look at the heavens with their solar, starry, and planetary glories; the sky with its clouds; the waters descending from above or wandering at our feet; the animals, the trees, the plants; and consider the permanency of their actions and conditions under the government of these laws. The most delicate flower, the tenderest insect, continues in its species through countless years, always varying, yet ever the same. When we think we have discovered a departure, as in the *Aphides, Medusæ,*

Distomæ, etc., the law concerned is itself the best means of instituting an investigation, and hitherto we have always found the witness to return to its original testimony. These frail things are never ceasing, never changing, evidence of the law's immutability. It would be well for a man who has an anomalous case before him, to contemplate a blade of grass, and when he has considered the numerous ceaseless, yet certain actions there located, and his inability to change the character of the least among them, to recur to his new subject, and, in place of accepting unwatched and unchecked results, to search for a like certainty and recurrence in the appearances and actions which belong to it.

In his peroration Faraday gives us the clearest view of his reasons for believing that scientific enquiry was among the highest and most demanding forms of mental education. To acquire 'judgement' is to arrive at conclusions objectively and carefully from the evidence available, increasing the volume of evidence where necessary by interrogating the world through carefully constructed experiment. Patience, effort and humility are the required attributes:

The education which I advocate will require *patience* and *labour of thought* in every exercise tending to improve the judgement. It matters not on what subject a person's mind is occupied, he should engage in it with the conviction that it will require mental labour. A powerful mind will be able to draw a conclusion more readily and more correctly than one of moderate character; but both will surpass themselves if they make an earnest, careful investigation, instead of a careless or prejudiced one; and education for this purpose is the more necessary for the latter, because the man of less ability may, through it, raise his rank and amend his position. I earnestly urge this point of self-education, for I believe it to be more or less in the power of every man greatly to improve his judgement. I do not think that one has the complete capacity for judgement which another is naturally without. I am of opinion that all may judge, and that we only need to declare on every side the conviction that mental education is wanting, and lead men to see that through it they hold, in a large degree, their welfare and their character in their own hands, to cause in future years an abundant development of right judgement in every class.

This education has for its first and its last step *humility*. It can commence only because of a conviction of deficiency, and if we are not disheartened under the growing revelations which it will make, that conviction will become stronger unto the end. But the humility will be founded, not on comparison of ourselves with the imperfect standards around us, but on the increase of that internal knowledge which alone can make us aware of our internal wants. The first step in correction is to learn our deficiencies, and having learned them, the next step is almost complete: for no man who has discovered that his judgement is hasty, or illogical, or imperfect, would go on with the same degree of haste, or irrationality, or presumption as before. I do not mean that all would at once be cured of bad mental habits, but I think better of human nature than to believe that a man in any rank of life, who has arrived at the consciousness of such a condition, would deny his common sense, and still judge and act as before. And though such self-schooling must continue to the end of life to supply an experience of deficiency rather than of attainment, still there is abundant stimulus to excite any man to perseverance. What he has lost are things imaginary, not real; what he gains are riches before unknown to him, yet invaluable; and though he may think more humbly of his own character, he will find himself at every step of his progress more sought for than before, more trusted with responsibility and held in pre-eminence by his equals, and more highly valued by those who he himself will esteem worthy of approbation.

Given his views on mental education, and the paramount role played by science in developing it, it is not surprising that Faraday was asked to give evidence to a Royal Commission in 1862 enquiring into the conduct and content of education. His responses to the commissioners' questions about the place of the natural sciences in mental development, compared with studies in classics and mathematics are trenchant:

> The phrase 'training of the mind' has to me a very indefinite meaning. I would like a profound scholar to indicate to me what he understands by the training of the mind; in a literary sense, including mathematics. What is their effect on the mind? What is the kind of result that is called the training of the mind? Or what does the mind learn by that training? It learns things, I have no doubt. By the very act of study it learns to be attentive, to be persevering, to be logical. But does it learn that training of the mind which enables a man to give a reason in natural things for an effect which happens from certain causes? Or why in any emergency or event he does (or should do) this, that, or the other? It does not suggest the least thing in these matters. It is the highly educated man that we find coming to us again and again, and asking the most simple question in chemistry or mechanics; and when we speak of such things as the conservation of force, the permanency of matter, and the unchangeability of the laws of nature, they are far from comprehending them, though they have relation to us in every action of our lives. Many of these instructed persons are as far from having the power of judging these things as if their minds had never been trained.

As a supremely successful example of a person who had been largely self-educated, Faraday was firm in proposing that young minds should be opened to the wonders and subtlety of scientific knowledge and concludes his evidence to the commissioners:

> I am not an educated man, according to the usual phrase-

ology and, therefore can make no comparison between languages and natural knowledge, except as regards the utility of language in conveying thoughts. But that the natural knowledge which has been given to the world in such abundance during the last 50 years should remain untouched, and that no sufficient attempt should be made to convey it to the young mind growing up and obtaining its first views of these things, is to me a matter so strange that I find it difficult to understand. Though I think I see the opposition breaking away, it is yet a very hard one to overcome. That it ought to be overcome, I have not the least doubt in the world.

Chapter 13

Final Days

As the 1860s approached, Faraday's memory (never good, as his earlier letters tell) deteriorated still further. Much of his writing throughout the final decade of his life is dominated by the frustration of not remembering how to end what he had begun, as in a letter of 1857 to Reverend John Barlow, who at the time was the Honorary Secretary of the Royal Institution. Nevertheless, Faraday analysed his defect with undiminished clarity:

> I am in town, and at work more or less every day. My memory wearies me greatly in working, for I cannot remember from day to day the conclusions I come to, and all has to be thought out many times over. To write it down gives no assistance, for what is written down is itself forgotten. It is only by very slow degrees that this state of mental muddiness can be wrought either through or under. Nevertheless, I know that to work somewhat is far better than to stand still, even if nothing comes of it. It is better for the mind itself—not being quite sure whether I shall ever end the research, and yet being sure, that if in my former state of memory, I could work it out in a week or two to a successful and affirmative result.
>
> Do not be amazed by what I am telling you. It is simply the thing I remember to tell you. If other things occurred to my mind, I would tell you of them. But one thing which often

Faraday at the age of 72.

withholds me is that if I begin a thing, I find I do not report
it correctly, and so naturally withdraw from attempting it.
One result of short memory is coming curiously into play
with me. I forget how to spell. I dare say if I were to read
this letter again, I should find four or five words of which
I am doubtful ('withholds, wearies, successful,' etc.) but I
cannot stop for them, or to look to a dictionary (for I had
better cease to write altogether), but I just send them, with
all their imperfections, knowing that you will receive them
kindly.

To more intimate friends, such as Miss Harriet Moore,
who painted numerous watercolours of the interior of Faraday's
laboratory and flat at the Royal Institution, he let thoughts of
approaching mortality obtrude. Even then, a side remark about

Miss Harriet Moore, the painter of the watercolours shown on pages 68, 107, 112, 114 and 122.

the American Civil War (the latter dates from 1861) shows a sharp (even waspish) observation of contemporary affairs:

> I have been writing to you (in imagination) during a full week, and the things I had to talk about were so many that I considered I should at last want a sheet of foolscap for the purpose; but as the thoughts rose they sank again, and oblivion covers all. And so it is in most things with me. The past is gone, *not* to be remembered, the future is coming, *not* to be imagined or guessed at. The present only is shaped to my mind. But, remember, I speak only of temporal and material things. Of higher matters, I trust the past, present, and future, are *one* with me, and that the temporal things may well wait for their future development.
>
> As with you, so with us. The harvest is a continual joy: all seems so prosperous and happy. What a contrast such a state is to that of our friends the Americans, for notwithstanding all their blustering and arrogance, selfwilledness, and

nonsense, I cannot help but feel drawn towards them by their affinity to us. The whole nation seems to me as a little impetuous, ignorant, headstrong child under punishment, and getting a little sobering experience, quite necessary for its future existence as a decent well-behaved nation amongst nations.

To his old friend Auguste de la Rive, Professor of Physics at Geneva University, and a correspondent for 30 years, Faraday's religious faith came most strongly to the surface: hope for a life to come comforted his final years:

I cannot tell when I wrote you last. Of late years I have kept a note, but I suppose I have forgotten to note. Having no science to talk to you about, a motive, which was very strong in former times, is now wanting. But your last letter reminds me of *another motive*, which I hope is stronger than science with both of us, and that is the future life which lies before us. I am, I hope, very thankful that in the withdrawal of the powers and things of this life, the good hope is left with me, which makes the contemplation of death a comfort—not a fear. Such peace is alone in the gift of God, and as it is He who gives it, why should we be afraid? His unspeakable gift in His beloved Son is the ground of no doubtful hope, and *there* is the rest for those who (like you and me) are drawing near the latter end of our terms here below. I do not know, however, why I should join you with me in years. I forget your age. But this I know (and feel as well) that next Sabbath day (the 22nd) I shall complete my 70th year. I can hardly think myself so old as I write to you—so much of cheerful spirit, ease and general health is left to me, and if my memory fails, why it causes that I forget *troubles* as well as pleasure and the end is, I am happy and content.

Faraday had ever been a staunch royalist and supporter of the monarchy. A letter from Queen Victoria's Private Secretary, Sir James Clark, elicited the following reply in February 1863:

I would, if I might, express my reverence for the Queen, the wife and the mother whose image dwells in the hearts of all her people. I wish that I were, as a subject, more worthy of her. But the vessel wears out, and at seventy one has but little promise for the future. The fifty years of use in the Royal Institution has given me wonderful advantages in learning, many friends, and many opportunities of making my gratitude known to them, but they have taken the matter of life and, above all, memory out of me, leaving the mere residue of the man that has been, and now I remain in the house useless as to further exertion, excused from all duty, very content and happy in my mind, clothed with kindness by all, and honoured by my Queen.

On 11 October 1861 Faraday addressed the following letter to the managers of the Royal Institution, who had been his employers for the past 48 years, and under whose aegis he had carried out the whole of his life's scientific work. At the age of 70 he resigned his post:

Gentlemen,

It is with the deepest feeling that I address you.

I entered the Royal Institution in March 1813, nearly forty-nine years ago and, with the exception of a comparatively short period during which I was absent on the continent with Sir Humphry Davy, have been with you ever since.

During that time I have been most happy in your kindness, and in the fostering care which the Royal Institution has bestowed upon me. I am very thankful to you, and your predecessors, for the unswerving encouragement and support which you have given me during that period. My life has been a happy one and all I desired. During its progress I have tried to make a fitting return for it to the Royal Institution and through it to Science.

But the progress of years (now amounting in number to three score and ten) having brought forth, first, the period

Faraday with Professor Brande.

of development, and then that of maturity, has ultimately produced for me that of gentle decay. This has taken place in such a manner as to render the evening of life a blessing, for whilst increasing physical weakness occurs, a full share of health free from pain is granted with it; and whilst memory and certain other faculties of the mind diminish, my good spirits and cheerfulness do not diminish with them.

Still I am not able to do as I have done. I am not competent to perform, as I wish, the delightful duty of teaching in the Theatre of the Royal Institution, and I now ask you (in consideration for me) to accept my resignation of the *Juvenile Lectures*. Being unwilling to give up what has always been so kindly received and so pleasant to myself, I have tried the faculties essential for their delivery, and I know that I ought to retreat, for the attempts to realize (in the trials) the necessary points brings with it weariness, giddiness, fear of failure and the full conviction that it is time to retire. I desire therefore to lay down this duty; and I may truly say that, such has been the pleasure of the occupation to me, that my regret must be greater than yours need or can be.

And this reminds me that I ought to place in your hands the *whole* of my occupation. It is, no doubt, true that the Juvenile Lectures, not being included in my engagement as Professor, were, when delivered by me, undertaken as an extra duty, and remunerated by an extra payment. The duty of research, superintendence of the house, and other services still remains. But I may well believe that the natural change which incapacitates me from lecturing, may also make me unfit for some of these. In such respects, however, I will leave you to judge, and to say whether it is your wish that I should still remain as part of the Royal Institution.

I am Gentlemen, with all my heart;

Your faithful and devoted Servant

M. FARADAY

Two years later we find him writing to the German chemist Justus Liebig that he was no longer engaged in philosophy (or as we would say, science):

I will write no more philosophy or have any ability as an active philosophic mind. It is the past that moves me, the remembrance of all the dear thoughts and associations that I have in former years been permitted to share in; and though I may have (through wear and years) to give up the race and fall into the rear, yet it rejoices me to think that those who still run carry forward a kind remembrance of me.

There cannot be any doubt that those who still run, and continue the enterprise of scientific enquiry, most certainly do

Faraday's retirement house at Hampton Court: in a nineteenth-century engraving (top) and today (bottom).

carry forward a kind remembrance of this remarkable human being.

The last recorded writing by Michael Faraday is a letter to his old friend Carl Friedrich Schoenbein (see Chapter 5) on 18 September 1862 which, because it is so brief and so affecting in its affliction, is reproduced here in its entirety:

DEAR SCHOENBEIN

Again and again I tear up my letters, for I write nonsense. I cannot spell or write a line continuously. Whether I shall recover—this confusion—do not know.

I will not write any more. My love to you

ever affectionately yours

M. FARADAY

Schoenbein described it to Faraday's biographer Bence Jones as 'hardly written, in a trembling hand'. One can easily imagine how it was that Schoenbein felt he could not bring himself to answer it though, as he put it 'the interruption of our epistolatory intercourse grieved me to the innermost of my heart'.

Michael Faraday died on 25 August 1867, sitting quietly in his armchair, at his home, a Grace and Favour house at Hampton Court given to him by Queen Victoria. He was buried in the Sandemanian plot in Highgate Cemetery under a small plain stone that records only his name and the dates of his birth and death.

Bibliography

[1] James F A L J (ed) *The Correspondence of Michael Faraday* (London: Institution of Electrical Engineers) 1991 vol 1 (1811–1831); 1993 vol 2 (1832–1840); 1996 vol 3 (1841–1848)

[2] Bence Jones H 1870 *The Life and Letters of Faraday* 2 vols (London)

[3] Williams L P 1965 *Michael Faraday: A Biography* (New York: Basic Books); reprinted by Da Capo Press Inc., New York, 1987

[4] Williams L P (ed) 1971 *The Selected Correspondence of Michael Faraday* 3 vols (Cambridge: Cambridge University Press)

[5] Martin T (ed) 1932–1936 *Faraday's Diary* 7 vols (London: G Bell)

[6] Bowers B and Symons L (ed) 1991 *Curiosity Perfectly Satisfyed: Faraday's Travels in Europe 1813–1815* (London: Peter Peregrinus)

[7] Faraday M 1861 *The Chemical History of a Candle* (London: John Murray)

[8] Faraday M 1878 *On the Various Forces of Nature and Their Relations to Each Other* ed W Crookes (London: Chatto and Windus)

[9] Jeffreys A E 1960 *Michael Faraday: A List of His Lectures and Printed Writings* (London: Chapman and Hall)

[10] Cantor G 1991 *Michael Faraday: Sandemanian and Scientist* (London: Methuen)

Acknowledgments

Whilst the majority of the illustrations are taken from material at the Royal Institution, my warmest thanks are due to the following for permission to reproduce works in their possession: Mrs M J St Clair (cover photograph); The Institution of Electrical Engineers (for the figures on pages 22 and 49); Glasgow University Library (for the figure on page 197).

Index

N.B.: Because it would otherwise appear in almost every entry in the index, the name Faraday has been used sparingly and should be understood in most entries.